Direct Numerical Simulations - An Introduction and Applications

Edited by Srinivasa Rao

Published in London, United Kingdom

IntechOpen

Supporting open minds since 2005

Direct Numerical Simulations - An Introduction and Applications
http://dx.doi.org/10.5772/intechopen.84930
Edited by Srinivasa Rao

Contributors
Myron Polatayko, Paul Medwell, Ian James Shaw, Jordan Kildare, Michael Evans, Alfonso Chinnici, Ciaran Andrew Sparks, S N Hossain, Rey Chin, Hussein A. M. Al-Zubaidi, Scott A. Wells, Isabelle Sochet, Kevin Gault, Latih Habeeb, Riyadh Sabah Saleh Al-Turaihi, Vasily Biryukov, Lucien Veleva, Dipal Patel, Martin Agelin-Chaab, Srinivasa P Rao, Luc Hakenholz. Gloria Acosta, Luis Chávez and Juan L. López

First published in London, United Kingdom, 2021 by IntechOpen
IntechOpen is the global imprint of INTECHOPEN LIMITED, registered in England and Wales, registration number: 11086078, 5 Princes Gate Court, London, SW7 2QJ, United Kingdom
Printed in Croatia

British Library Cataloguing-in-Publication Data
A catalogue record for this book is available from the British Library

Additional hard and PDF copies can be obtained from orders@intechopen.com

Direct Numerical Simulations - An Introduction and Applications
Edited by Srinivasa Rao
p. cm.
Print ISBN 978-1-83880-558-6
Online ISBN 978-1-83880-559-3
eBook (PDF) ISBN 978-1-78984-836-6

Meet the editor

Dr. P. Srinivasa Rao is presently working as a Professor and Head of the Department of Mechanical Engineering at Vardhaman College Engineering, Hyderabad. Dr. Rao earned his Ph.D. in Computational Fluid Dynamics (CFD) and IC Engine Combustion. He worked as a scientist in the Computational Fluid Dynamics division of the Scientific Engineering and Computing Group (SECG) at the Centre for Development of Advanced Computing (C-DAC), Pune. Dr. Rao has over 60 publications in refereed international journals and conferences and is an inventor of 4 patents. He acted as a facilitator for the learning process and organized 16 workshops/FDPs/SDPs, 3 international conferences, and 2 national conferences beneficial to faculty, researchers and industry and delivered 32 plenaries, keynote speeches, and invited talks. Dr. Rao has taught over 32 courses on CFD, turbulence modeling, and combustion. He has advised 4 doctoral research fellows and has been a research visitor for 39 graduate students. He has been the editor of international scientific journals and a reviewer for more than twenty journals from Asia, Europe, and the USA of repute like SAE. Dr. Rao's research has involved applications of the CFD to the problems of mechanical and aerospace systems, computational physics, and turbulence modeling including biology and medicine.

Contents

Preface

Problems in fluid mechanics and turbulence are omnipresent and pose challenges to engineers, physicists, and mathematicians alike. Particularly, the flow instabilities and flow transition from laminar flow to a chaotic or turbulent state, lead to changes in the properties of the fluid across space and time. The only alternative is a numerical solution, and this is why computational methods have become an indispensable tool in the exploration and analysis of a wide range of engineering and physical phenomena. Direct Numerical Simulation (DNS) is an advanced branch of computational fluid dynamics (CFD) devoted to the high-fidelity solution of turbulent flows.

The strength of DNS is the ability to provide complete knowledge, unaffected by approximations, at every point and every time of the simulation within the flow. Therefore, it is an ideal tool to address basic research questions regarding turbulence in engineering and physics. However, the cost of DNS prevents it from being used as a general-purpose design tool. Though there are large amounts of information available on this subject in the literary world and on the web, there are insufficient resources that reference matching case studies. This book is not a systematic journey through DNS but rather an exposition of doing it. It is more of an explicit example-based problem-solving guide by various authors in specialized numerical experimentation. The target audience consists of engineering students venturing into computational fluid dynamics. This book is an introductory illustrative text on DNS and may help readers in gaining fundamental skills on the techniques of direct numerical simulations in computational fluid mechanics.

I thank all the authors for contributing such wonderful content for the readers of this book. I thank the management of Vardhaman College of Engineering, for supporting me in editing this volume. I thank the principal, deans, heads, and my colleagues. Special thanks to my student Shashank for helping in a chapter. I thank my wife, beloved daughter, Amma, Nanagaru, Kavya and all my family members for their continuous support.

Dr. P. Srinivasa Rao,
Vardhaman College of Engineering,
Shamshabad, Hyderabad, Telangana State, India

Section 1

Inviscid Flow Turbulence

Chapter 1

Turbulent Flow Simulations

Laith Jaafer Habeeb and Riyadh Sabah Saleh Al-Turaihi

Abstract

This chapter consists of four sections; Introduction, Boundary and initial conditions, Setting of nonuniformly spaced grid, and Simulation approach. The fields of the fluctuating velocity describe the turbulent flows. Such fluctuations blend the transported quantities, like species concentration, energy, and momentum and make the transported quantity fluctuations in addition. Because such fluctuations have a small scale and high frequency, so they are very computationally costly for simulation straightforward in the virtual engineering computations. Alternatively, the instantaneous accurate governing equations can be time-averaged, ensemble-averaged, or in different way handled for removing the small scales, causing a modified group of equations, which are computationally less costly for solving.

Keywords: periodic, inflow, outflow and thermal field boundary conditions, initial conditions, grid system, nonuniformly spaced grid, numerical simulation, turbulent inflow conditions

1. Introduction

The updated evolution in technology of computer has offered a fresh insight into the turbulent flow phenomenon. Particularly, a numerical analysis approach named direct numerical simulation (DNS), which is clearly depended upon the physical laws with no turbulence model, is anticipated to become too beneficial due to the Navier–Stokes equation reliability and versatility. Nevertheless, the DNS used to the turbulent flows with intricate shapes of boundary or large Reynolds (Re) nos. Puts a great challenge to the engineers, who are doing different efforts for overcoming the relevant problems. The grids that are staggered have traditionally been utilized for the computation, and the grids that are non-staggered are more often utilized in the recent time. For example, investigations on the organized grid systems have progressed quickly owing to their expansion to the generalized coordinate regimes. The use of DNS to the fields of the turbulent flow with the free surfaces, like seas and rivers, was initially tried by Lam and Banerjee in 1992. From that time, such method has supplied significant information on the turbulent structures and mechanism of the mass transport close to the free surfaces that are hard for investigating experimentally. The whole of such investigations have utilized either finite difference method or spectrum method utilizing a staggered grid, and there have been no try for using a normal grid for a method.

2. Boundary and initial conditions

2.1 Periodic boundary conditions

In the hydrodynamically and thermally fully (time) developed conditions as well as for the periodically fully developed conditions, where the flow uniformity in the stream-wise direction is possible to be assumed, the periodic boundary conditions (BCs) can be employed on the whole variables along the stream-wise direction, involving temperature and pressure. The channel flow is the highly often utilized geometries in the DNS, where the periodic BCs are entailed. The turbulent flow simulations in a channel having an infinite width (z) and length (x) are conducted via approximating the domain infinite extent in the span-wise and stream-wise directions with a finite length computational domain (named unit cell) and via using the periodic BCs in such directions. These circumstances theoretically impose the depending variable fields to become the periodic along the chosen direction. Practically, their use relies upon the specific approach itself: for example, the periodicity is inherent to the spectral approaches' principle functions, whereas in the finite volume/finite difference methods, it is performed via entailing the computational stencils' circular topology at the domain double terminals. A particular care must be considered when selecting the unit cell extent that must be highly sufficient for encompassing the flow biggest scales. When such states are met, the flow inside the computational domain could be regarded as accurate flow depicting in the infinite channel.

2.2 Inflow open boundary conditions

The periodic BCs could be used in the stream-wise direction to calculate variables if fully developed circumstances are satisfied. In case of the attention is a spatially evolving flow, the approach of DNS needs re-sorting to a method of the inflow turbulence creation for the fields of temperature and velocity. Recently, the methods of the inflow turbulence creation for the field of velocity have been categorized and surveyed by Wu in 2017. A short subsidiary periodic domain, where the turbulence is calculated in parallel to the principal computation, is giving outcome results utilized as inflow state for target simulation. Such kind of inflow (BCs) is recognized as a re-cycling approach. Fabricated approaches represent a substitute trial for fully modeling the inflow.

2.3 Outflow open boundary conditions

Contriving proper BCs at the open boundaries if the average flow exists, the domain can be a hard work. The highly familiar difficulties of the outflow open BCs are that they do not make genuine reflection of the velocity and of the transported scalars and they try to make wiggles at the boundary domain and might influence the solution algorithm stability. Then, the outflow BCs have to depict the physical domain continuation after calculating the domain edge, and the transport equations extrapolated from the domain interior are frequently enforced in this form:

$$\frac{\partial \phi}{\partial t} + U_c \frac{\partial \phi}{\partial x} = 0 \tag{1}$$

In several situations, a consistent value for U_c that ensures the mass conservation is utilized. The interpretation for the convective velocity U_c can be entailed in Eq. (1) and depended upon the configuration specific flow. For instance, some researchers, who dealt with the turbulent jets and plumes and proposed using the exponential function for $U_c(y)$. In the survey on the open outflow circumstances by the approach was detailed as a promising method for the buoyant, turbulent plumes. The approach suggested by some authors was depended upon a 1-D advection diffusion expression at the outflow, where the term of diffusion was added to Eq. (1) and a phase velocity was presented as well as U_c.

2.4 Boundary conditions for the thermal field

The geometry of turbulent channel is also highly often utilized for the investigations of the heat transfer of near-wall. The highly exact method is a conjugate heat transfer, where the heat conduction within the actual heated walls is considered. The related investigation for the uses of liquid metal is the DNS that conducted a conjugate heat transfer (DNS) at low Prandtl no. (Pr = 0.01). Majority of studies of (DNS) heat transfer were conducted with the simple thermal BCs with no solid walls, and dimensionless temperature at the contact plane of fluid–solid was set to zero:

$$\theta(y = \pm h) = 0 \text{ (non} - \text{fluctuating thermal BC)} \tag{2}$$

Such kind of thermal BC is a too good approximation of actuality if the ratio of thermal activity $K = \sqrt{(\lambda\rho c_p)_f/(\lambda\rho c_p)_w}$ moves to zero, whereas the thickness of the heated wall and parameter $(G = \alpha_f/\alpha_w)$ stay finite$(\alpha = \lambda/\rho c_p)$. In such situation, the temperature fluctuations created in fluid are not penetrating into solid wall. Such kind of perfect thermal BC was represented as a non-fluctuating BC. A case study of this system is the solid and fluid air/metal combination. For the combination of water/steel, thermal BC is yet near to the non-fluctuating BC when the heated wall is sufficiently thick; nevertheless, this approximation does not succeed if the thickness of metal wall is small. The thermal BC that allows the peak penetration of the turbulent temperature fluctuations from fluid to solid is expressed as.

$$\langle\theta(y = \pm h)\rangle_{x,z,t} = 0 \quad \left(\frac{\partial\theta'}{\partial y}\right)_{y=\pm h} = 0 \text{(fluctuating thermal BC)} \tag{3}$$

The average dimensionless temperature for fluctuating temperature (BC) at wall is yet set to zero, whereas the zero derivative is suggested for the temperature (θ') fluctuating component. Such kind of thermal BC is performed if the ratio (K) of thermal activity moves to infinity; experiments can be repeated with a water flume, which is heated with a too thin metal foil. In 2001, when initial analysis of BCs (Eqs. 2 and 3) in a turbulent flow of channel was conducted by Tiselj et al. (2001b), such BCs were represented as "isothermal" BC (Eq. 2) and "isoflux BC for a dimensionless temperature" (Eq. 3). Particularly, the "isoflux" term results much confusion, since it was often blended with the standard "isoflux" BC for a physical temperature. Such pair of BCs must not be wrong: the aspect with a fixed volumetric heat origin within the solid walls matches to the "isoflux" BC for the physical temperature. The explained BC in such situation can be either non-fluctuating (Eq. 2) or fluctuating (Eq. 3). Hence, Tiselj and Cizelj [1] recommended new names for BCs (Eqs. 2 and 3). A significant characteristic of various thermal BCs (Eqs. 2 and 3) is that in the passive scalar approximation, they possess approximately negligible effect on the coefficient of heat transfer. In other words, the average temperature profiles close to the wall are specially the explained thermal

(BC) independent. Certainly, an important discrepancy can be noted in the temperature fluctuations and the other turbulent statistics. For actual fluid–solid regimes, the thermal BC is common within the two bounded BCs of Eqs. (2) and (3). For the composition of steel and liquid sodium with (K) equals (1), the thermal (BC) is approximately in the middle within the non-fluctuating and fluctuating temperature (BC). Thus, so as to investigate the temperature fluctuation details with their penetration inside the solid wall, the method of a conjugate heat transfer is advised for the heat transfer (DNS) in the liquid metals.

2.5 The initial circumstances

Beyond the whole equations, computational domain and BCs have been identified; the initial circumstances are needed for beginning the simulations. Provided that the statistics for a long term are not dependent of the initial circumstances and given (CPU) time is not a bounding factor, it is frequently enough to begin with any type of primary velocity containing in minimum some perturbations of long enough wavelengths. Since such simulations' initial stages are run on the highly rough grids, the consumption of CPU time should not become very high. In major situations, if the geometry stays alike, it is enough to utilize instantaneous fields determined from the simulations at the other Re or Pr nos. Beyond proper scaling. In state of computational expense is a bounding factor; thus the primary transient is shortened with a few more advanced kinds of the initial circumstances.

3. Setting of nonuniformly spaced grid

3.1 The grid system outline

The generation of computational grid can be divided into two types: a regular grid regime and a staggered grid regime in accordance with the evaluation points' arrangement of pressure and velocity. The staggered types have been widely utilized in the spatial separation approaches of the analysis of flow either in the MAC or Simple approach since they govern the solution numerical vibration. Nevertheless, they need many interpolating processes, and calculating the loads is timorous, particularly if a generalized coordinate regime is employed. So, the grid regime choice is tightly related to the turbulent flow computation compactness [2]. On the other hand, the regular grids have been indicated the suffering from the taking place of solution spatial vibration. Nevertheless, they are suitable for the programming since the physical quantities are specified at similar point. The other advantages of these grids are the properness to improve the solution accuracy and the ease of conversion into a generalized coordinate regime that is influential in the deal with the intricate boundary difficulties. The regarded numerical vibration can be eliminated via the nonuniformly spaced grid adoption, and the active simulation can be recognized via utilizing a rough grid over the areas where the velocity spatial variations are small, with exception of the near-wall zone.

3.2 Influence of nonuniformly spaced grid

Because the nonuniformly spaced grid adoption, in general, results in lowered accuracy, here the influences of such kind on the numerical solution are debated from a 1-D linear convection problem referred by Eq. (4):

$$\frac{\partial u}{\partial t} + U\frac{\partial u}{\partial x} = 0, \quad (U = 1, -\infty < x < \infty)u$$
$$(x, 0) = \begin{cases} \dfrac{1 - \cos(2x - \pi)}{2}, & \left(\dfrac{\pi}{2} \le x \le \dfrac{3\pi}{2}\right) \\ 0, & otherwise \end{cases} \tag{4}$$

As depicted in **Figure 1**, three kinds of grids are employed: a uniform-spaced grid, a smooth nonuniformly spaced grid (A) having a continuous metric (x_ζ), and a smooth nonuniformly spaced grid (B) having discontinuities in metric (x_ζ). Because the accurate solution of Eq. (4) is really similar to the primary waveform and does not vary with the time, the numerical solution beyond the (m-th) cycle is compared with the accurate solution via the periodic boundary condition adoption for the two terminals of x-axis. The scheme of the third-order Adams-Bashforth is employed for the time progress.

The comparison of waveforms beyond 50 cycles of computation for a Courant no. of 0.1 in state of the uniform-spaced grid is presented in **Figure 2**. Graphs from (a) to (c) at the top depended upon a central difference scheme, while the graphs from (d) to (f) at the bottom side are based on an upwind difference scheme. This figure manifests that the nonuniformly spaced grid (A) causes somewhat a higher attenuation error than that for the uniform-spaced grid. That is likely resulted via the truth that the accuracy is reduced by one order if the grid spacing does not smoothly vary in the nonuniformly spaced grid, as it can be concluded from the second-order differential approximation truncation error which is inferred via the Taylor expansion. Graphs from (c) to (f) propose that such reduction of accuracy can be substituted via the higher-order finite difference approximation adoption. As for the nonuniformly spaced grid (B), not merely the error of attenuation but also the phase one is created in the whole situations. Since such a grid possesses similar smoothness as the nonuniformly spaced grid (A), the error of phase is regarded to be associated with the metric (x_ζ'S) discontinuities. Concerning the finite difference convection term scheme, the upwind difference is, in general, stated as a sum of the numerical viscosity and the central difference. So, the central differences in the graphs from (a) to (c) be the upwind differences in the graphs from (d) to (f) via the addition of truncation errors (numerical viscosity) of the second-, fourth-, and sixth-order differential, correspondingly. Therefore, the numerical viscosity

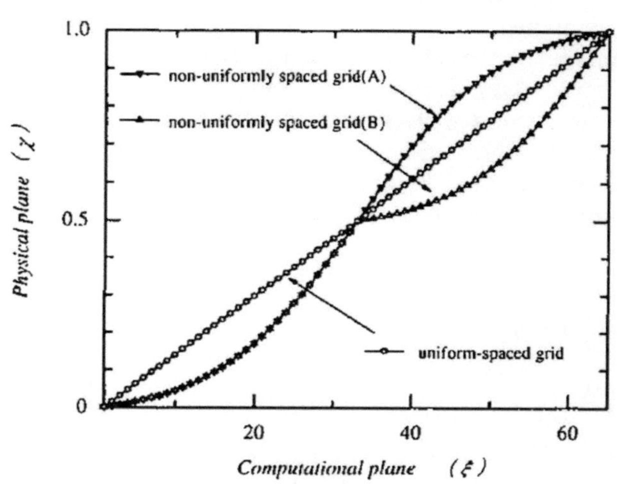

Figure 1.
Coordinate transformation.

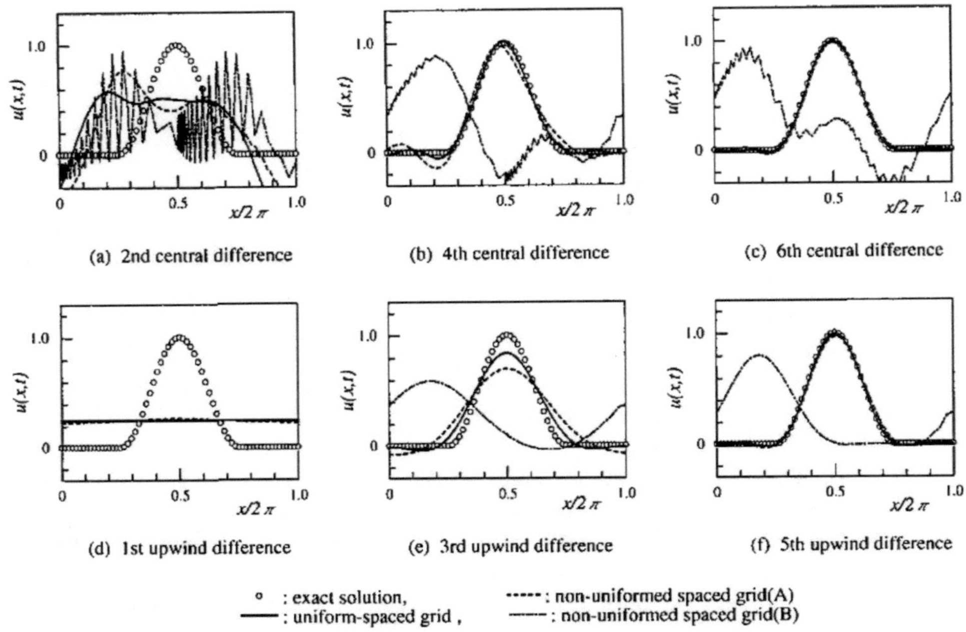

Figure 2.
Waveform comparison for different types of grid spacing ($c = 0.1, x/2\pi = 50$).

influence can be studied via comparing the graphs at the top of **Figure 2** with those at the bottom. While the numerical viscosity possesses a suppressing influence of the vibrating solution obtained in the central difference, the influence is highly deep in the lower-order upwind differences so as the solution is skewed. Even the third-order upwind difference that is frequently utilized in the practical calculation is not fitted for the analysis of turbulence in which too high accuracy is needed. The fifth-order upwind difference that entails a small reverse influence on the solution works as a filter with a high cut, which eliminates merely the components having a high wavenumber. In the central differences, even in state of high-order accuracy, the nonlinear instability is probably triggered via the dispersion which resulted via the truncation errors of the odd-numbered differential. So, it is regarded that the fifth-order upwind difference adoption is active in eliminating the aliasing errors that are resulted via the components having a high wavenumber created in a continuous way from the nonlinear terms and in the suppressing solutions, which include the numerical vibrations having no physical meaning.

Therefore, in the current analysis of the study utilizing a regular grid in a generalized coordinate regime, a nonuniformly spaced grid with enough smoothness and a continuous metric is created. Also, the scheme of the fifth-order upwind difference is employed for the term of convection.

4. The approach of simulation

4.1 Numerical aspects

The DNS of the turbulent boundary layer)TBL(with the embedded instruments, like vortex generators, represents a step in intricacy [3], and it is likely with the NTS code [Owing to its combination of numerical dissipation, multi-block capability, and implicit numeric]. This code supplies the ability for the accurate computations with time with the structured multi-block overset grids and uses the scheme of an implicit second order in time flux-difference splitting. The flow is dealt as an incompressible. The inviscid fluxes in governing equations are approximated with the centered fourth-order accurate differences everywhere, with the exception of the devices' stagnation points near vicinity where the third-order upwind approximations are utilized. Which is required for suppressing the odd–even pressure oscillations resulted via the vigorous gradients of pressure in such areas. The viscous fluxes are approximated with the centered second-order differences. At each step of time, the resulted finite-difference equations are solved with the employment of Gauss-Seidel relaxations via planes and sub-iterations in "pseudo-time."

4.2 Turbulent inflow conditions

It is the more original aspect of the approach. It is preferred to possess a standard TBL that encounters the device, which controls the flow, without a long method zone. Some researchers revealed that the inflow conditions which depended on the random nos. Can simply take 20δ for recovery; this would be too wasteful if the beneficial zone is of the order of 25δ. Their method of recycling method, in some sense, wasted around 8δ that is too smaller and could likely have been decreased more with no damage. Here, the method is vigorously inspired via LWS, but it is simpler, and a number of separate enhancements were done. The LWS used a various processing to the outer and inner zones of TBL and mixed the two expressions. That may seem optimum, but it needs re-scaling the velocity of friction as well as the thickness of BL and presents many arbitrary parameters and almost intricate expressions. This method failed to have the benefits of a pair of desirable facts. The first is that the near-wall turbulence re-creates itself too quickly than the outer-zone turbulence, and so a small damage is made via using the outer-zone scaling in every part. The second is that if the recycling plant is becoming too near the inflow that is favorable in respect to the cost of computing, the clashing between the outer-zone and the inner-zone scaling basically disappears. Also, a short recycling distance decreases the shifting in the span-wise length scales and the time scales, neither of which is taken into consideration (so, the recycling is better modified with shorter distances, till the distance is too short that the recycling and inflow planes are too near that no eddies development can occur). For such causes, the current method utilizes a single re-scaling. In a similar way, the corrections to the component (v) of the wall-normal velocity possess a too little influence and are deleted. Eventually, there is no subsidiary simulation here. The initial simulation yields its own recycling information.

The method conclude of utilizing the velocity field at a "recycling" plant ($x = x_r$), which is partway within the domain, properly changed, to give the velocity field at the inflow ($x = 0$). Because the BL is somewhat thicker at (x_r), the scaling in the wall-normal direction (y) is the minimum change, which is required. Let the thicknesses at ($x = 0$) and ($x = x_r$) be (δ_o) and (δ_r). Thus, the inflow condition for the velocity vector (U) is:

$$U(0, y, z, t) = U(x_r, y\delta_r/\delta_0, z + \Delta z, t - \Delta t) \tag{5}$$

where the span-wise shift, Δz, is presented to preserve the turbulence at the inlet and the recycling sections out of the phase (in the simulations debated below, the Δz value was put equal to half period of the span-wise). That is a small step, which is taken for disorganizing any lasting span-wise changes of the average flow that would otherwise be recycled and likely possess more time for damping via the span-wise diffusion. Utilizing the condition from the final step of ($t - \Delta t$) is highly suitable.

The procedure simplicity is visible in comparison with the LWS, but the fundamental concept is theirs. The generalization to flows with the gradients of pressure and the curved boundaries stays to be done. The operation is governed via a pair of two parameters; the difference of the (Re) no. from the inflow to recycling, $\Delta R_x = x_r U_e / v$, and (δ_0/δ_r) ratio. By assuming the approximate scaling for the thickness of BL, the flow should be stabilized with the following magnitude for the thickness of inflow:

$$\delta = 0.37\left(\frac{v}{U_e}\right)\cdot\left[\frac{\Delta R_s}{(\delta_r/\delta_0)^{5/4} - 1}\right]^{4/5} \tag{6}$$

This provides a fine control over the thickness of inflow (despite the factor (0.37) is not accurate, the little discrepancy in the rates of growth between the thickness of momentum s and the thickness of BL is ignored, and the turbulence is not the exact evolved condition), in which the response to the little variations in (δRx) and (δ_0/δ_r) can be predicated (as the flow is properly set). The primary conditions for this simulation have some of significance. For example, the random perturbations, which are very feeble or very insufficient in length scales, can too well "die out." When it is, the simulation is becoming laminar and producing a higher thickness of BL than the anticipated with the turbulence, due to the various scaling than in the above equation, which will make the obvious failure. That is more possibly with a short distance of re-cycling. The (δR_x) and (δ_0/δ_r) little values costly affect the steady state but resist the turbulence maturation, which is primarily required. So, it can be useful to begin with the recycling plant farther downstream and then move it nearer to the inflow via decreasing (δR_x) and (δ_0/δ_r) in agreement. Instead of, when a device, which controls the flow, finishes the normal BL and bounds the recycling plant location, the simulation can begin with a further length that is attached to the upstream grid, which would later be removed beyond the turbulence is well established, for reducing the long simulation's cost required for an adequate sample time. Moreover, the early part of the simulation does not require the complete length of the zone containing the recovery and device, particularly when the parameters of the recycling are being modified. In brief, the flexibility in re-beginning the simulations from each other with various domains will be too useful, and many approaches exist for surmounting the problems of the early system, based upon the preference of the user. As the procedure of recycling is properly setup, the solution loses the memory of the primary state processing. The method used here takes the advantages from another code, so as it is not a universal. It is as the following: a RANS solution is computed with the wanted thickness, and the perturbations is added to it. Such perturbations are determined from a code, which is utilized for initializing the uniform turbulence simulations. It gives turbulence volume of cubes shape with 0.063 size, each one has random phases that are positioned side-by-side upon the wall beyond their velocity field is multiplied via a shape function, so as they become zero at the wall and in the free stream. That acts properly, despite nothing is made for injecting the exact Reynolds shear stress. An appropriate solution is determined in some flow-through periods of times.

Author details

Laith Jaafer Habeeb[1*] and Riyadh Sabah Saleh Al-Turaihi[2]

1 Training and Workshop Center, University of Technology, Baghdad, IRAQ

2 Department of Mechanical Engineering, College of Engineering, Babylon University, Babil, Iraq

*Address all correspondence to: laithhabeeb1974@gmail.com; 20021@uotechnology.edu.iq

References

[1] Tiselj I, Stalio E, Angeli D, Oder J. Direct numerical simulations for liquid metal applications. In: Thermal Hydraulics Aspects of Liquid Metal Cooled Nuclear Reactors. 2019. pp. 219-244

[2] Hayashi S, Ohmoto T, Koreeda N, Takikawa K. Study on turbulent structure of open channel flow by direct numerical simulation using regular grid system. In: Computational Mechanics - New Frontiers for New Millennium.

[3] Spalart PR, Strelets M, Travin A. Direct numerical simulation of large-eddy-break-up devices in a boundary layer. International Journal of Heat and Fluid Flow. 2006;27:902-910

Chapter 2

Propagation of Shock Waves in Two Rooms Communicating through an Opening

Isabelle Sochet, Kevin Gault and Luc Hakenholz

Abstract

Confined explosions represent a serious safety hazard as significant damage to humans and structures is observed, unlike in free-field explosions. An experimental small-scale study investigated the blast wave in a single-story building. The blast waves were generated by the detonation of a gaseous charge. The building was divided into two rooms by a movable wall which could be positioned at three different locations. The presence of an opening in this movable wall means that two rooms were considered: a transmitter room (TR) and a receptor room (RR). The configuration without the movable wall was also studied. Pressure profiles recorded with pressure gauges at ground level and on the wall presented numerous reflections. The damage effects were severe since the maximum overpressure never fell below 0.2 bar. Although this study is limited to a small scale and gaseous detonation charge, the results can be applied to a large scale and for a TNT charge.

Keywords: blast wave, confined explosion, explosion effects, overpressure, pressure profile, reflection, scaling

1. Introduction

Nowadays, safety in public, industrial, or military areas is a major concern. In spite of constant improvements in rules and standards, the risks linked to accidental or intentional explosions in industry are still significant. The numerous explosions reported demonstrate the importance of increasing the protection of people and structures in open, semi-confined, and full confined environments. Recent examples are the explosion in 2000 of a firework workshop (Enschede, the Netherlands) [1], in 2013 at the West Fertilizer Company (Texas, USA) [2], in 2015 a harbor warehouse (Tianjin, China) [3], and in 2016 the explosion in a manufacture of basic chemicals, fertilizers and nitrogen compounds, plastics, and synthetic rubber in primary forms (Coatzacoalcos, Mexico) [4]. A long list of terrorist attacks can be added to these industrial accidents. The most recent bombing attacks occurred on 13 May 2018, when a series of bombs exploded in three churches in Surabaya (Indonesia); on 1 July 2018, when a suicide bomber detonated a bomb in Jalalabad (Afghanistan); and on Easter Sunday 2019 when several bombs exploded in the capital of Sri Lanka, Colombo.

Numerous studies have been carried out in recent years to analyze the behavior of shock waves in the air [5–8]. While analytical, empirical, and numerical studies have been conducted to predict the overpressure generated by the detonation of a solid or gaseous mixture in air, few studies have been done in closed environments, and most of the tools and models developed in air cannot be used in closed spaces because of the complex phenomenology of a confined detonation. Nevertheless, one can find in the literature several studies on semi-confined [9] or urban [10] configurations. Confined explosions present a serious safety hazard as significant damage to humans and structures is observed, unlike in free-field explosions.

Recent studies in a full confined environment are sparse, however, due to several limitations. In numerical studies the modeling of a full confined environment can become extremely costly in terms of computational resources. Researchers have therefore developed 1D/3D hybrid models to achieve a good balance between the accuracy of the results and the time required [11]. Experimentally, it is complex and costly to set up a full-scale experiment to study the propagation of the shock wave inside a building. As a result, studies on confined detonations are made at small scale. Reichenbach et al. [12] pointed out the interest of scaled experiments.

The advantages of small-scale experiments are their good flexibility and low cost. Thanks to the Hopkinson similarity law and the studies by Baker [13], the results recorded at small scale can be extrapolated to full scale if the charges are the same type between the two scales. Ohrt et al. [14] studied the propagation of a shock wave in a 1/12 small-scale model built from the full-scale model. The detonation of pressed TNT and a Composition A5 (98.5% RDX and 1.5% stearic acid) was studied and visualized in the small-scale model. A good correlation was found between the results at small scale and full scale.

Ram et al. [15] considered a single-story building at a 1:100 scale. Different configurations inside the building were investigated, all with an opening on the front face and inside. The shock wave was produced outside, impacted the front face, and propagated inside. The authors compared the internal geometry to a stiff porous building and depicted the pressure profile by a low-pass filter. They demonstrated that the simulation of full-scale experiments reproduced the scaled-up experiments.

The afterburning consequences on the shock wave reflection in confined spaces have also been investigated. The additional energy released by the secondary mechanism of combustion increases the energy in the whole flow field and increases the pressure. Therefore, the reflections interact with the post-combustion products. A quasi-static pressure is recorded in the case of a pyrotechnic explosive. Experimental and numerical studies have focused on the afterburning effect, for example, the work of Togashi et al. [16] and Milne et al. [17].

Hazard zones in confined rooms have been identified with a dual approach (experimental and numerical). Massoni et al. [18] investigated a three-level building at small scale and carried out numerical simulations. In this work the shock wave was generated by a shock tube placed in the vicinity of the model. Miura's team [19, 20] evaluated safety in nuclear fuel cycle facilities. The model comprised three rooms at ground level and two rooms on the second floor. Strong pressures were recorded on the corners and after diffraction around doors and windows. These studies showed that the corners of rooms are a critical zone.

Another relevant area of application for confined explosion studies is tunnels. For example, the experiments carried out by Binggeli et al. [21] and the simulations presented by Rigas et al. [22] and Benselama et al. [11] showed that the velocity and pressure of the blast wave are reinforced in long narrow geometries. The presence of an orthogonal pipe reduces the overpressure and is a good solution to ensure human safety.

In addition to their flexibility, small-scale experiments can reliably predict the behavior of shock waves in a confined area. When the building geometries are complex, it is difficult to study and predict the behavior of the waves. Sauvan et al. [23] conducted an experimental characterization of the reflections coming from the detonation of a gaseous charge inside a cubic model. To obtain the most accurate characterization, the authors started with the free field and built the model wall by wall, making it possible to identify the first four reflected shocks.

The small-scale experiments conducted by Julien et al. [24, 25] on the behavior of shock waves coming from the detonation of a gaseous charge inside a small-scale warehouse led to the creation of predictive laws for the main characteristics of the shock wave. They also showed that critical zones exist in complex confined buildings.

Although confined explosions have been researched, the effects of the size of the opening between two rooms have not been fully established.

The present work concerns the volume and open area effect on the shock wave propagation inside a single-story building with an inner movable wall. The research is based on a small-scale experiment. The study focuses on the impact of the open area and of the room volume on the shock wave main pressure profiles and maximum overpressure.

This chapter describes the results obtained from these experiments. Pressure distribution is discussed and potential damage is highlighted.

2. Experimental setup

The experimental setup is a small-scale closed one-story building with a movable wall that can divide the building into two rooms: a transmitter room (TR) where the explosion (CE) takes place and a receptor room (RR). The walls are made of medium-density fiberboard (**Figure 1**). The internal dimensions of the building are length (L) 1.23 m, width (Lw) 0.51 m, and height (h) 0.24 m. All the walls have the same thickness e = 0.04 m. It means that the exterior dimensions are 1.31 m for the length, 0.59 m for the width, and 0.28 m for the height.

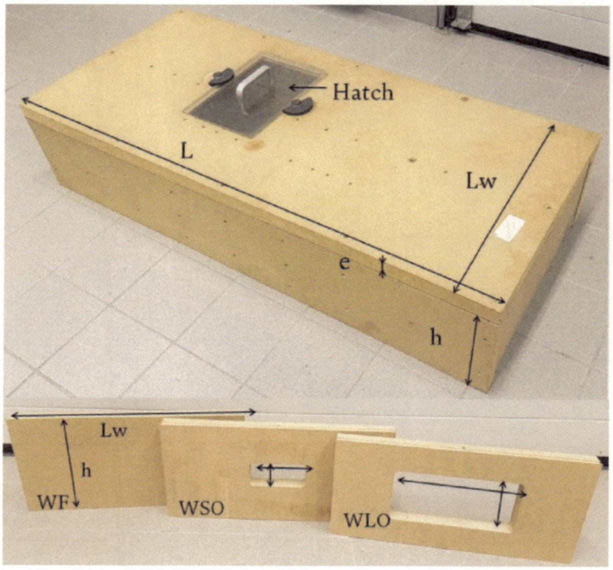

Figure 1.
Test facility.

The dimensions of the movable walls are Lw = 0.51 m and h = 0.24 m. Two types of movable wall were considered: a full wall (WF) and walls with an open area.

The openings are centered on the face. The wall with the smallest opening is denoted WSO, and the dimensions of its opening are 0.128 $\times$ 0.06 m, i.e., an area of 0.00768 m^2. The wall with the largest opening is denoted WLO, and the opening has an area of 0.0305 m^2 (0.254 $\times$ 0.12 m). The movable wall can be fixed at different locations inside the building to vary the volumes of the transmitter and receptor rooms.

Four configurations were examined (**Figure 2**). Configuration 1 is without the movable wall, configuration 2 has a movable wall located 0.275 m from the south wall, and configurations 3 and 4 have a movable wall located 0.594 m and 0.909

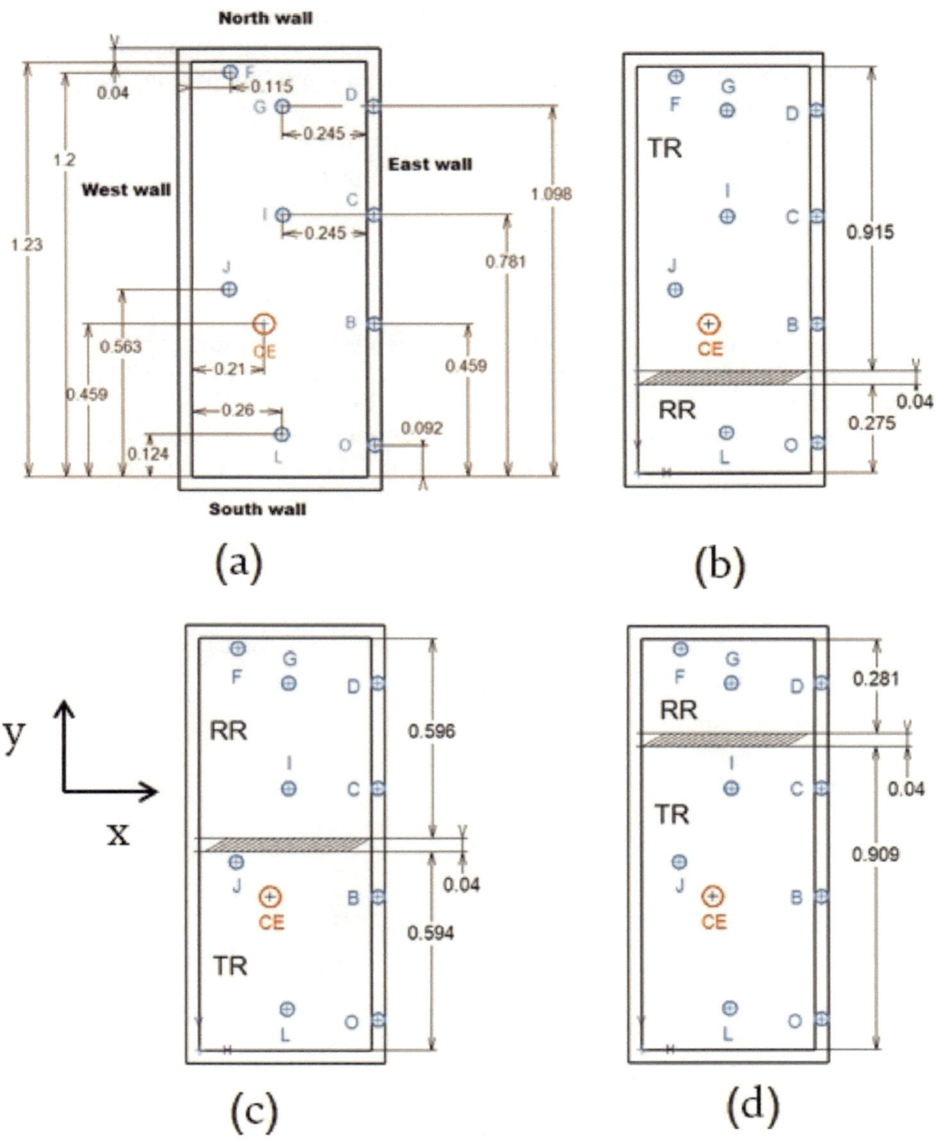

Figure 2.
Schematic experimental configurations with pressure gauges and explosive charge (CE) positioning: (a) configuration 1, (b) configuration 2, (c) configuration 3, and (d) configuration 4.

from the south wall, respectively. In the text, we use the terms configurations 2, 3, and 4 or walls 2, 3, and 4.

The explosive charge is a stoichiometric propane oxygen ($C_3H_8 + 5O_2$) mixture confined in a hemispherical soap bubble with a radius of 0.07 m. The experiments were conducted at ambient temperature and pressure. The density of the gaseous mixture is 1.41 kg m^{-3}, so the mass of explosive charge is 1.0129×10^{-3} kg.

The detonation of the gaseous mixture is generated by an exploding wire fixed between two electrodes linked to a high-voltage supply. The exploding wire is transformed into a plasma and delivers severe conditions to initiate the detonation (all details can be found in [26]). The center of the explosive charge (CE) is fixed (**Figure 2**) for all the configurations studied. In C1 the distance between the south wall and the center of explosion is 0.459 and 0.771 m from the north wall. The distance between the movable wall and the center of explosion is 0.144 m in C2, 0.135 m in C3, and 0.450 m in C4. The volume of the transmitter room (TR) in C2 is $V_{TR\text{-}C2} = 0.1120$ m^3, in C3 is $V_{TR\text{-}C3} = 0.0727$ m^3, and in C4 is $V_{TR\text{-}C4} = 0.1113$ m^3. So, the volume of receptor room (RR) is $V_{RR\text{-}C2} = 0.0337$ m^3, in C3 is $V_{RR\text{-}C3} = 0.0729$ m^3, and in C4 is $V_{RR\text{-}C4} = 0.0344$ m^3.

Pressure histories were recorded over a 6 ms period with nine pressure transducers (Kistler 603B) at an acquisition frequency of 1 MHz. Five transducers (L, J, I, G, F) were distributed on the ground and four (O, B, C, D) on the east wall at half height (0.12 m). A layout of the sensor distribution with the different configurations (depending on the position of the movable wall) can be found in **Figure 2**. A sensor is considered protected when the shock wave cannot hit it directly.

3. Analytical approach to reflections

The pressure profile analysis requires knowing the type of reflection produced inside the single-story building. A preliminary study was carried out in order to identify whether there was a transition from regular reflection to Mach reflection and, if so, the height of the Mach stem. This transition between the two types of reflection is defined by a maximum angle β_{max} of oblique reflection which allows the formation of a Mach stem. For an incident shock at Mach 2.8, the limit angle β_{max} is 39.23°, and for a more intense incident shock, this angle reaches 39.97° [7].

The transition distance of formation R_{i0} is defined by [7]:

$$R_{i0} = \text{HOB}.\tan(\beta_{max}) \quad (1)$$

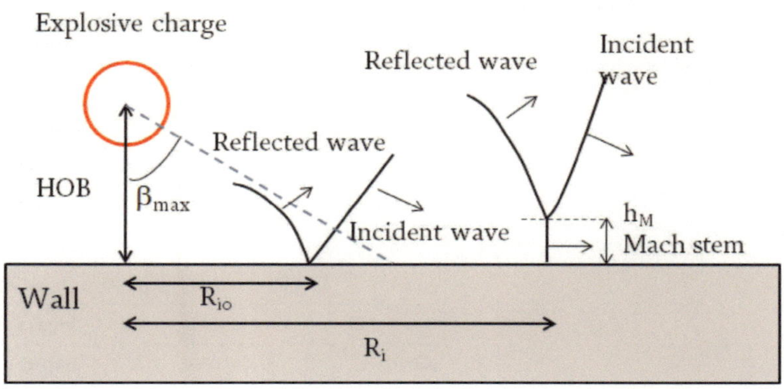

Figure 3.
Scheme of Mach stem formation.

and the height of the Mach stem h_M by:

$$h_M = 0.07\ \mathrm{HOB}\left(\frac{R_i}{R_{i0}} - 1\right)^2 \tag{2}$$

where HOB represents the height of burst, i.e., the distance of the explosive charge center to the wall, and R_i is the distance from the wall (**Figure 3**).

On this basis, the distance R_{i0} and Mach stem height were calculated for each configuration by considering $\beta_{max} = 39.23°$. Only scenarios for which the Mach stem is received by a pressure gauge or by a wall are reported in **Table 1**.

Configuration	Wall reflection (WR)	R_{i0} (m)	Interest point	h_M at interest point (IP) (m)	Comments
1	East	0.245	L	3×10^{-3}	$h_M < d_{WR\text{-}IP}$
			I	2×10^{-3}	$h_M < d_{WR\text{-}IP}$
			G	0.054	$h_M < d_{WR\text{-}IP}$
	West	0.171	L	0.010	$h_M < d_{WR\text{-}IP}$
			I	0.010	$h_M < d_{WR\text{-}IP}$
			G	0.110	$h_M < d_{WR\text{-}IP}$
			F	0.160	$h_M > d_{WR\text{-}IP}$
2	East	0.245	I	2×10^{-3}	$h_M < d_{WR\text{-}IP}$
			G	0.054	$h_M < d_{WR\text{-}IP}$
			F	0.086	$h_M < d_{WR\text{-}IP}$
	West	0.171	I	0.011	$h_M < d_{WR\text{-}IP}$
			G	0.109	$h_M < d_{WR\text{-}IP}$
			F	0.162	$h_M > d_{WR\text{-}IP}$
	South	0.117	East wall	0.024	Impact on wall
			West wall	6×10^{-3}	Impact on wall
3	East	0.245	L	3×10^{-3}	$h_M < d_{WR\text{-}IP}$
			South wall	0.016	Impact on wall
	West	0.171	L	0.013	$h_M < d_{WR\text{-}IP}$
			South wall	0.041	Impact on wall
	North	0.110	East wall	0.028	Impact on wall
			West wall	8×10^{-3}	Impact on wall
4	East	0.245	L	3×10^{-3}	$h_M < d_{WR\text{-}IP}$
			I	2×10^{-3}	$h_M < d_{WR\text{-}IP}$
			South wall	0.016	Impact on wall
			North wall	0.015	Impact on wall
	West	0.171	L	0.013	$h_M < d_{WR\text{-}IP}$
			I	0.011	$h_M < d_{WR\text{-}IP}$
			South wall	0.041	Impact on wall
			North wall	0.039	Impact on wall

Table 1.
Location of Mach stem.

There can be no Mach reflection on the south and north walls in configuration 1, as the walls are not wide enough to reach a transition of reflection. The transition appears on the east and west walls. On the east wall, the transition is at y = 0.25 m (south-north) from the CE. The height of the Mach stem is <1 cm at y = 0.322 m and reaches 0.05 m for $y = y_G = 0.639$ m. These heights are lower than the distance of the east wall to gauges I and G. However, the study of the reflection on the west wall leads to a transition distance of 0.17 m and a Mach stem height of 0.01 m for $y = y_I = 0.322$ m, 0.1 m for $y = y_G = 0.639$ m, and 0.16 m for $y = y_F = 0.741$ m. This implies that only gauge F is impacted by the Mach reflection.

In most cases, the results of the calculations reported in **Table 1** show that the height of the Mach stem does not exceed 0.1 m. Only gauge F can be impacted by the Mach stem in configurations 1 and 2 after reflection on the west wall. **Table 1** highlights that if a reflection is produced on a wall with a transition, then the Mach stem can reach a lateral wall. This is the case, for example, in configuration 3 with the reflection of the west wall. The Mach stem arrives on the south wall with a height of 0.04 m and a height of 0.028 on the east wall. This is also the case for configuration 4 if we consider the reflection on the west wall: the Mach stem impacts the south and north walls with a height of 0.04 m.

4. Pressure profile analysis

4.1 Effect of opening size on pressure profiles

The pressure profiles recorded in a confined building are complex due to the multiple reflections. In this part, the pressure signals are analyzed by considering the free field and C1 and the presence of the movable wall in C3 without and with an opening.

4.1.1 Gauges on the ground

4.1.1.1 Gauge J

In free field (**Figure 4a**), the secondary shock is detached from the incident peak at 0.297 ms with $\Delta P^{+} = 0.098$ bar. This secondary shock is amplified by the confinement and reaches 0.264 bar. After detonation of the explosive charge, a first shock wave propagates into the surrounding air, and a rarefaction wave propagates toward the center of charge. Once the rarefaction wave decreases the inside pressure, a new shock appears and propagates to the origin and reflects [27]. The effect of confinement in configuration 1 produces small reflection waves and consequently a small overpressure. The overpressure is limited to 0.2 bar with a peak at 0.5 bar.

The mobile wall P3 creates several reflections (**Figure 4b**) which are represented by the pressure peaks. Wall 3 is only 0.031 m away, so the second peak arrives immediately after the incident shock. The overpressure is increased to 3.6 bar due to the direct reflection.

The most severe peak appears later at 2 ms and reaches 4.7 bar. This peak decreases and arrives later with the presence of an opening in wall 3. The energy of the incident shock is not confined in the transmitter room but is partially transmitted to the receptor room. The smaller peak at 0.55 ms is stronger with the larger opening. This can be explained by the diffraction of the shock on the corner of the opening in the direction of gauge J. The corner of the larger opening is closer to gauge J than that of the smaller opening.

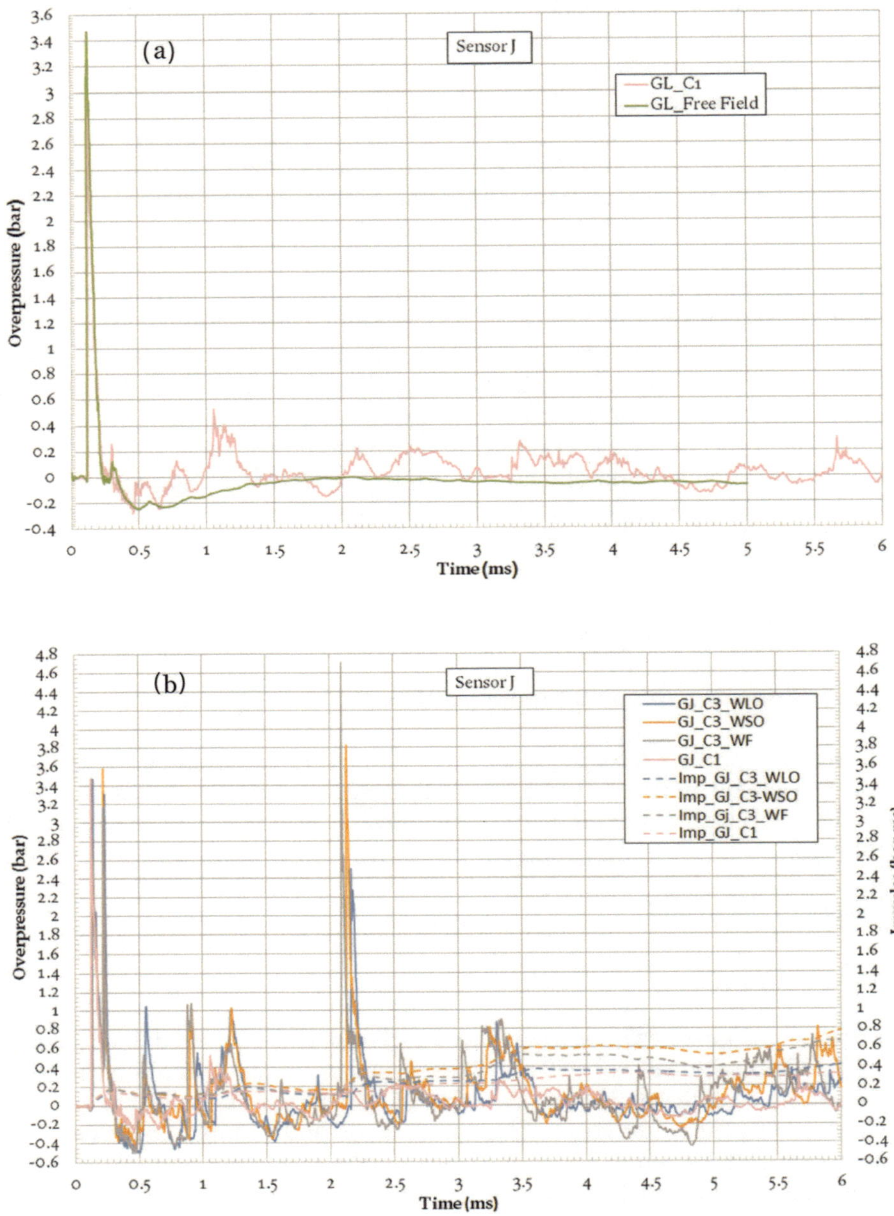

Figure 4.
Pressure and impulse histories. Gauge J: (a) comparison of free field and C1 and (b) configurations 1 and 3.

4.1.1.2 Gauge L

In the single confinement (C1), gauge L is closer to the south wall (0.124 m). In this configuration, the second peak is the maximum overpressure (**Figure 5a**) and comes from the south wall with a reflection coefficient on the order of 2. This is verified by applying the classical normal reflection equation [7]:

$$\frac{P_r}{P_i} = \frac{(3\gamma - 1)\frac{P_i}{P_0} - (\gamma - 1)}{(\gamma + 1) + (\gamma - 1)\frac{P_i}{P_0}} \tag{3}$$

where P_r and P_i are the absolute reflected and incident pressure and P_0 is the ambient pressure. In this case $P_i = \Delta P_i + P_0 = 1.794$ bar.

After that, the reflected shock interacts with the reflected wave from the west wall and the east wall. Strong negative pressures (−0.35 bar) are observed during an expanded time (1–5 ms) and provide an average impulse around 0.3 bar ms which reaches the value 0.44 ms.

The peak at 2.6 ms is not induced by the north wall because it is present in configuration 3 (**Figure 5b**). Wall 3 leads to new peaks at 2 and 4 ms. They correspond to a decreasing phase for configuration 1. They can emerge from the

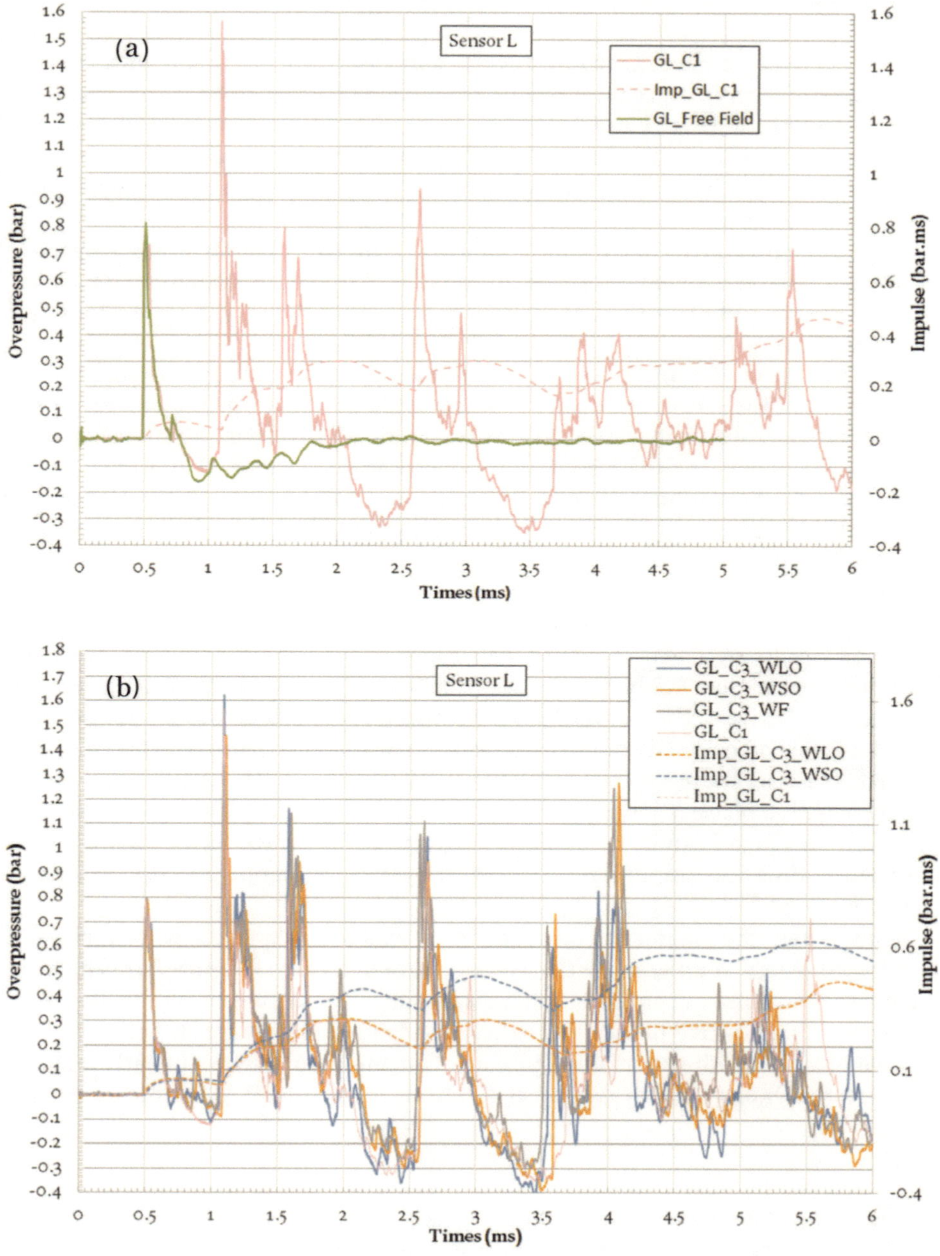

Figure 5.
Pressure and impulse histories. Gauge L: (a) comparison of free field and C1 and (b) configurations 1 and 3.

presence of a small Mach stem on the south wall coming from reflection on the west wall and another small Mach stem on the east wall coming from the north wall (**Table 1**). The pressure profiles versus the size of opening are very similar with a decrease in the overpressure with the increase in event size. A time offset appears after 3 ms with the presence of wall 3. A negative phase appears at 1.9 ms for the large opening and at 4.8 ms with the two openings. This is due to diffraction at the angles of the vent and consequently the generation of rarefaction waves. These rarefaction waves catch up the reflected waves and decrease them, as observed by Rose [28]. This phenomenon increases the impulse with the opening.

4.1.1.3 Gauge I

A strong overpressure is recorded at the same time (1.09 ms) and with an amplitude similar to the pressure profile from gauge L (**Figure 6a**). The overpressure is 1.595 bar for gauge I and 1.578 bar for gauge L. The peak cannot be explained by a unique reflection on the nearest wall (east wall) but by the cumulative effect of the reflected waves from the west and east walls. The reflection by the ceiling can be found at 1.25 ms.

After a negative phase, an increase in pressure up to 1.2 bar at 2.6 ms is observed and comparable to the peak recorded on gauge L. Three wave trains can be distinguished, corresponding to three time ranges: (1) 1.38–1.77 ms, (2) 3.5–4.12 ms, and (3) 4.8–5.56 ms.

Wall 3 (**Figure 6b**) mitigates the level of overpressure with a delay that increases when the size of the opening decreases. The larger opening decreases the overpressure at 2 ms from 4.7 bar to 2.54 bar (ratio of 1.85) with a time delay of 0.055 ms. Of course, the full wall 3 totally obstructs the propagation of the shock wave in the receptor room. The impulses are comparable in terms of evolution and magnitude between configuration 1 and configuration 3 with the larger opening.

4.1.1.4 Gauge G

Gauge G is located near the north wall. **Figure 7a** shows several peaks with an amplitude higher than 0.3 bar in the range of 6 ms. The ratio of the maximum overpressure between the first reflected wave and the incident wave is on the order of 2. However, on gauge G, the second peak is not the maximum overpressure reached. The maximum appears at 3.9 ms and is equal to 0.787 bar.

The negative phase is lower than that observed on gauge I. The wave train identified on gauge I can be recognized on gauge G. The wave train is more extended in time and amplified compared to the first reflection and the peak at 2.6 ms on gauge I.

As expected, the greater the obstruction by wall 3, the later the wave arrives in the receptor room and the more the overpressure decreases (**Figure 7b**). The negative phases resulting in configuration 1 disappear with the presence of wall 3 and correspond to a positive phase. Nevertheless, negative phases are still observed in the case of wall 3 with the smaller opening. The impulses increase linearly and are higher on this gauge than on gauge I: 0.4–0.5 bar ms against 0.35 bar ms for gauge I.

4.1.1.5 Gauge F

Gauge F is the closest gauge to the north wall. A wave train arrives almost simultaneously with the incident shock wave (**Figure 8a**).

The three wave trains identified on gauges I and G are still present on gauge F. On gauge F, there is an amplification of the overpressure during the first wave train

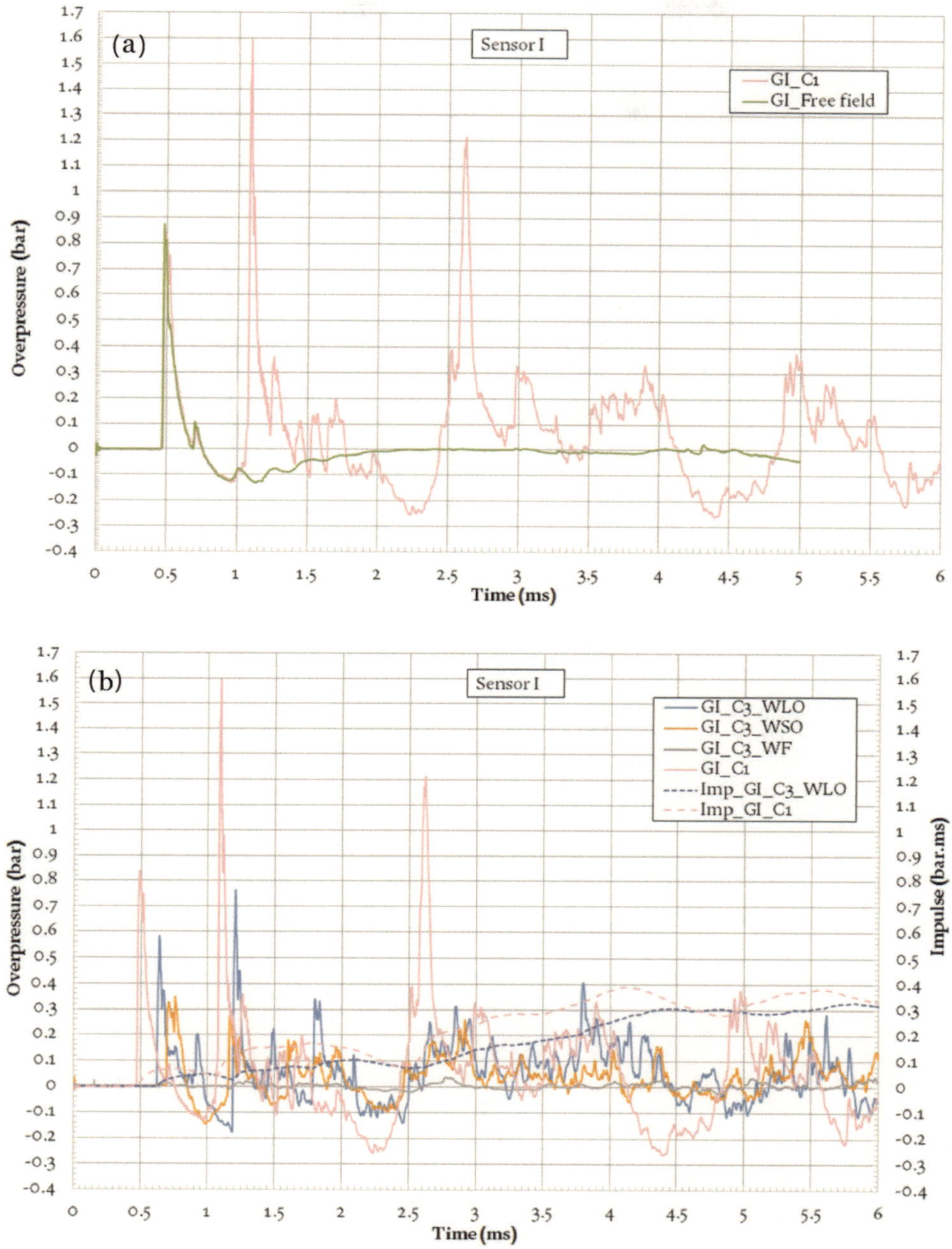

Figure 6.
Pressure and impulse histories. Gauge I: (a) comparison of free field and C1 and (b) configurations 1 and 3.

(1.54–2.6 ms), whereas the last one is mitigated compared to gauge G. This amplification can be explained by the presence of a Mach reflection identified previously (**Table 1**). The aggregation of the waves represents the interaction of the reflected waves coming from reflection on the four walls and the ceiling.

With wall 3 (**Figure 8b**), there is no negative phase. All the overpressures are strongly mitigated with the reduction in the opening size and totally mitigated with the full wall 3. However, a global view of the pressure profile shows that the same peaks can be identified without or with wall 3 with opening. This means that the reflected waves are produced in the second part of the building, i.e., in the receptor

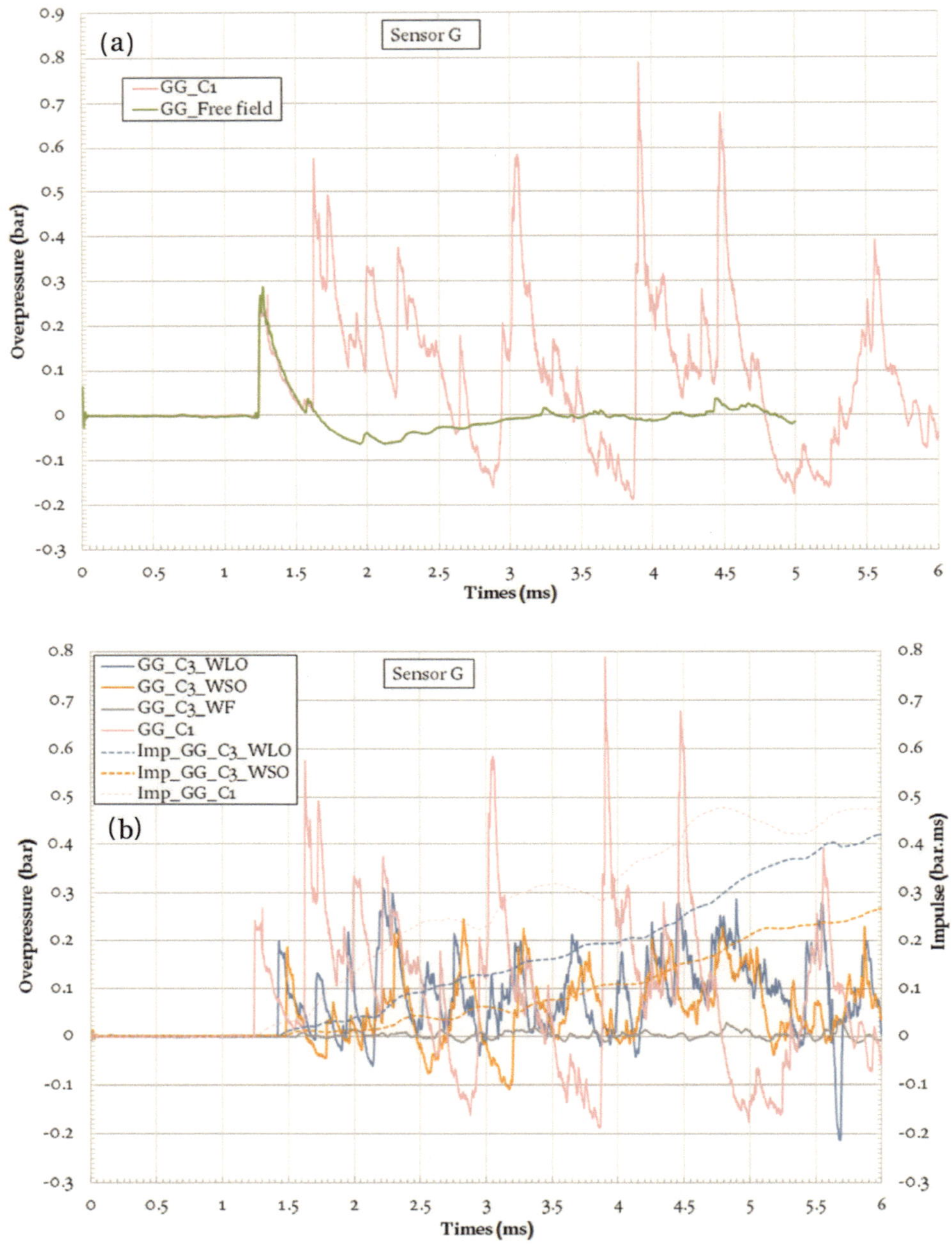

Figure 7.
Pressure and impulse histories. Gauge G: (a) comparison of free field and C1 and (b) configurations 1 and 3.

part. The impulses obtained on gauge F are similar to those recorded on gauge G in terms of evolution and magnitude.

4.1.2 Gauges on the east wall

Gauge B is directly impacted by the explosion (**Figure 9a**). This gauge undergoes the most severe overpressure (2.4 bar) but the lowest impulse compared to the other gauges. Negative phases recorded on gauge B correspond to positive

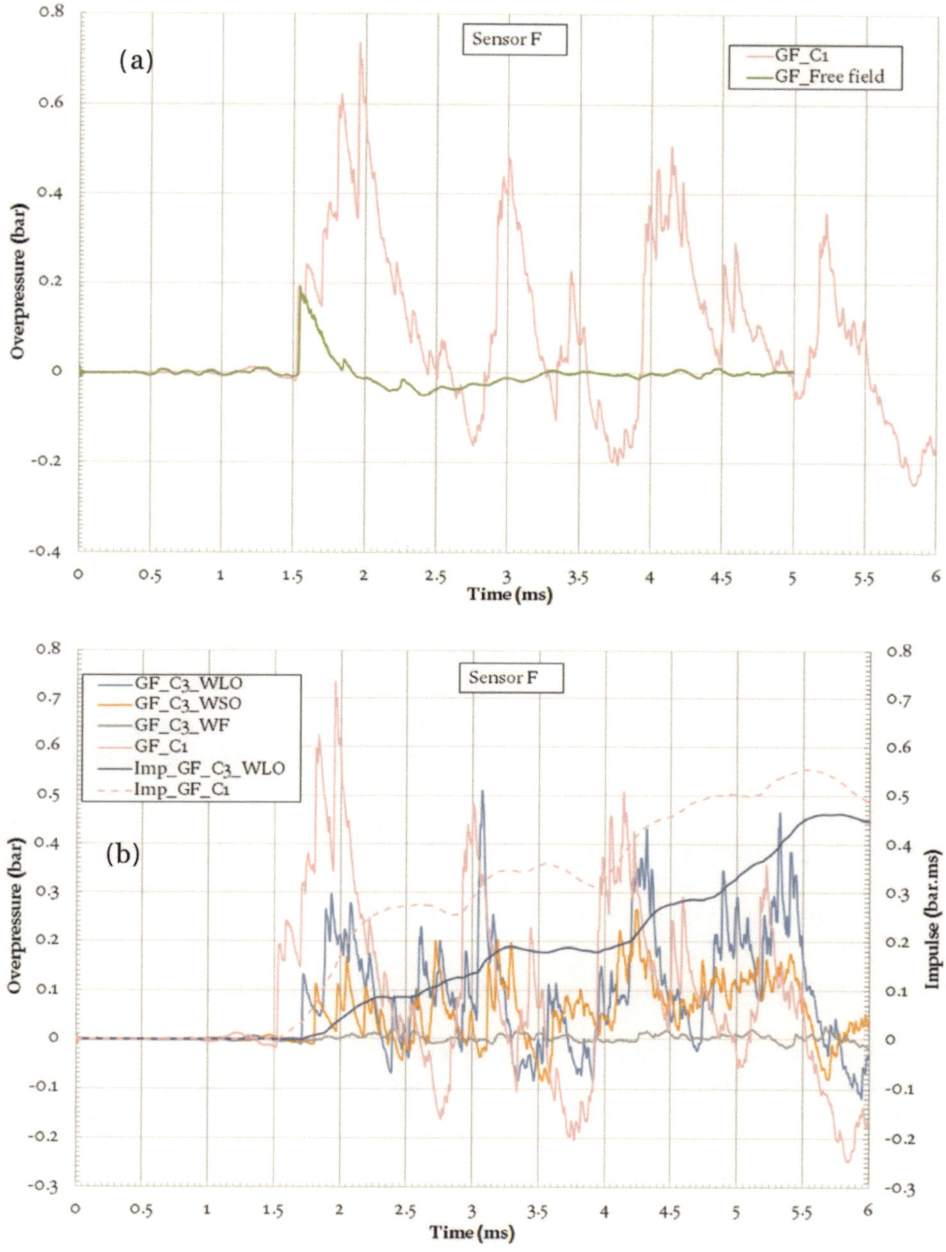

Figure 8.
Pressure and impulse histories. Gauge F: (a) comparison of free field and C1 and (b) configurations 1 and 3.

phases on gauge O (**Figure 9b**) and vice versa. The time delay of arrival on gauge O is not enough to explain the alternating phases.

The first two peaks until 1.3 ms observed on gauges C (**Figure 9c**) and O are superimposed. This is due to the almost identical distances of gauges O and C from the center of the explosive charge (0.489 and 0.456 m, respectively). Strong overpressure appears (0.9 bar, 1.5 ms) on gauge O due to the proximity of the south wall. On gauge C, a wave train with an overpressure above 0.3 bar is maintained between 2.8 and 3.7 ms, leading to a higher impulse level.

Gauge D receives the first wave later (1.4 ms), and the maximum of overpressure (1.2 bar) is reached at 3.5 ms (**Figure 9d**).

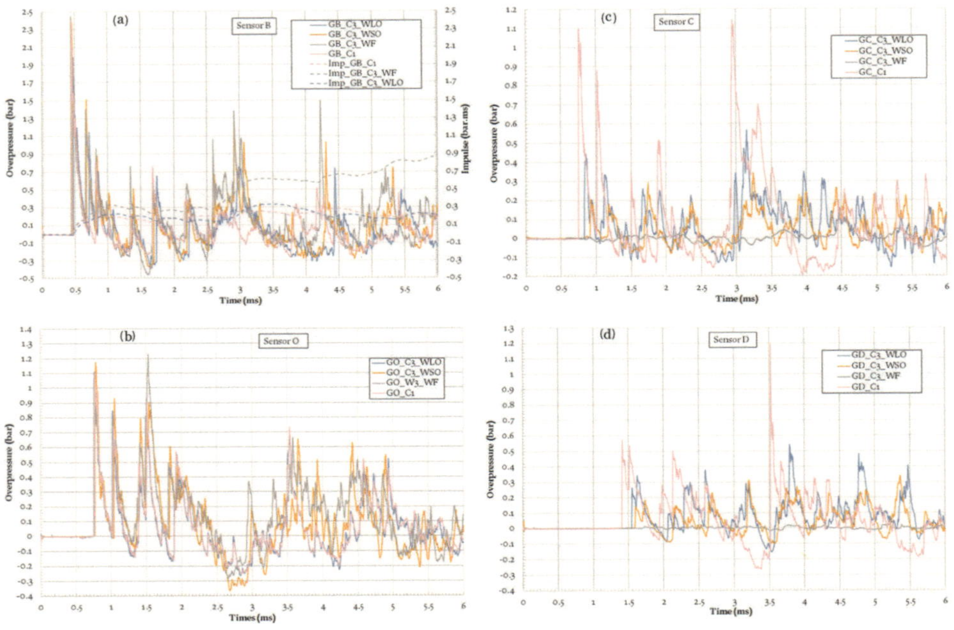

Figure 9.
Pressure and impulse histories. Configurations 1 and 3: (a) gauge B; (b) gauge O; (c) gauge C and (d) gauge D.

A similar evolution of pressure profiles is obtained on gauges I (**Figure 6b**) and C (**Figure 9c**) even if the waves arrive early. The same behavior is noted between gauges G and D (**Figure 7b** and **d**).

Wall 3 has little effect on the pressure profile of gauge O, except for a new peak with an overpressure of 0.37 bar at 2.95 ms. The overpressure decreases as the size of the opening increases. On gauges C and D, the overpressure increases with the size of the opening of the movable wall in configuration 3.

4.2 Effect of location of the movable wall on pressure profiles

In this section, the analysis concerns the effect of the volume in the transmitter room, i.e., the position of the movable wall inside the building. Here, we focus on gauges J and F and configurations with a full wall and with the larger opening of the wall.

4.2.1 Gauge J

In each configuration, gauge J is in the transmitter zone (**Figure 2**). Unfortunately, it is only possible to check the incident overpressure of the shock wave in configuration 4 due to a thermal drift (**Figure 10a**). However, the pressure profiles obtained in configurations 1 and 2 with the full wall are similar and different from configuration 3 examined in the previous section.

The presence of an opening in the mobile wall (**Figure 10b**) does not affect the response of gauge J in configurations 1, 2, and 4.

4.2.2 Gauge F

Gauge F is located near the north wall. So, this gauge is totally protected in configurations 3 and 4 with a full movable wall.

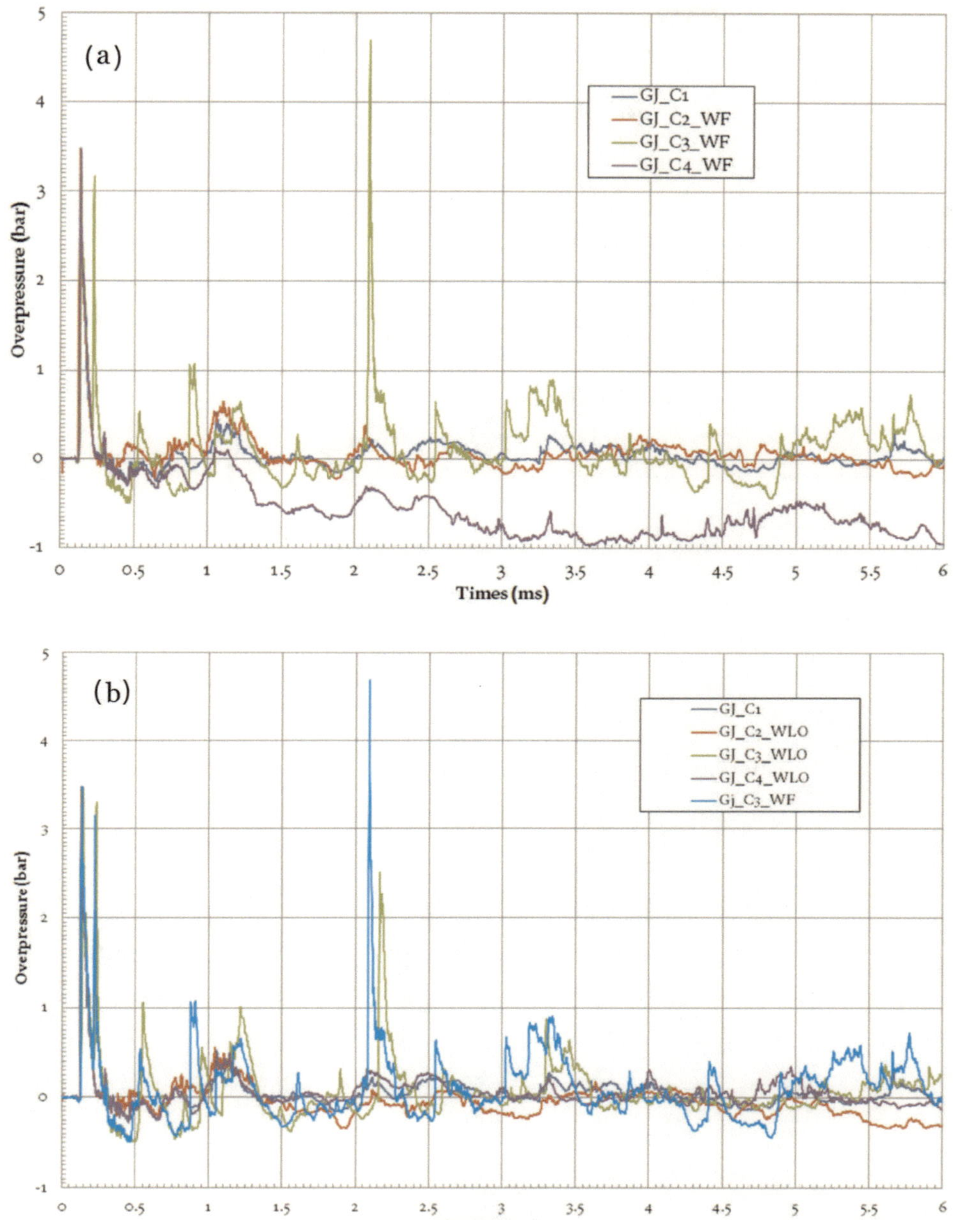

Figure 10.
Pressure histories. Gauge J, configurations 1–4: (a) with full wall and (b) with large opening for C2, C3, and C4.

The effect of the location of the movable wall is only perceptible above 2.2 ms. The wall deletes the negative phase existing before 3 ms in configuration 1 (**Figure 11a**). So, the overpressure is high in configuration 2. However, a series of three negative phases is recorded between 3.5 and 5.2 ms.

In the case of configurations with an opening in the mobile wall (**Figure 11b**), gauge F is not protected. The opening in configuration 2 mitigates all the pressure by filtering the peaks at 3.3 ms and 3.94 ms. In configurations 3 and 4, the negative phase is almost nonexistent. The more the volume of the receptor zone decreases, the more the overpressure signals are absorbed, falling to below 0.3 bar in configuration 4.

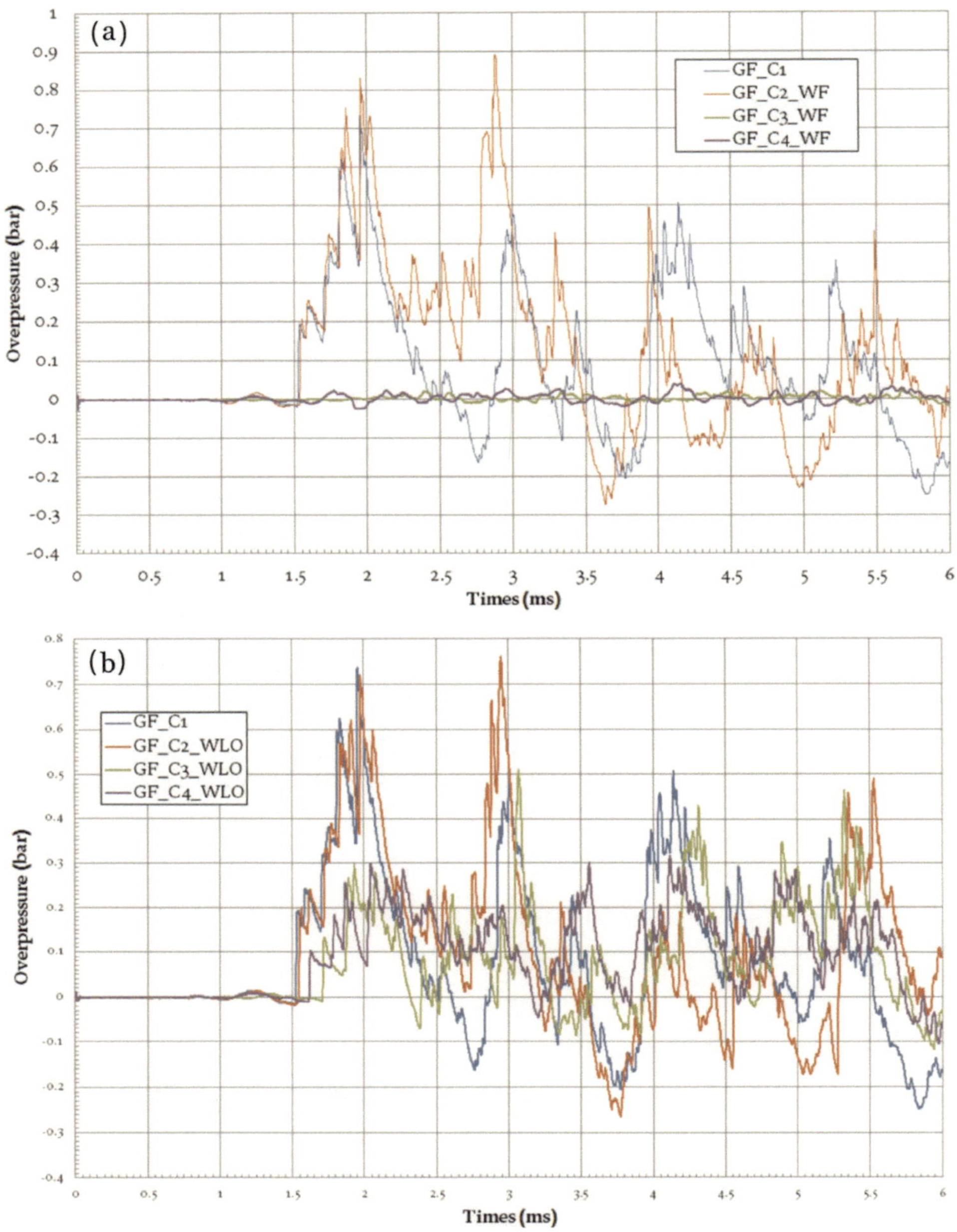

Figure 11.
Pressure histories. Gauge F configurations 1–4: (a) with full wall and (b) with large opening for C2, C3, and C4.

5. Mitigations and effect zones

Here we analyze the global pressure field and its effects inside the building. The data are expressed as a function of the scaled distance Z (m kg$^{-1/3}$) which is defined as $Z = R.M^{-1/3}$ where R is the radial distance from the charge center and M is the mass of the gaseous mixture. The evolution of overpressure in free field is represented by the following law [26]:

$$\mathrm{Ln}\frac{\Delta P}{P_0} = 1.98458 - 1.93917\ \ln Z + 0.100553\ (\ln Z)^2 \text{ with } 0.8 \le Z\left(\mathrm{m\ \ kg^{-1/3}}\right) \le 53.4 \qquad (4)$$

For confined geometries, the shortest path leading from the charge to the gauge position can be predicted by studying the reflections and diffractions encountered by the shock wave to reach the studied interest point (**Figure 12**).

Using this path, as illustrated in **Figure 12**, a set of confined scaled distances Z_{conf} is calculated by the following methodology. The shortest path gives a new set of radial distances between the center of the explosion and the sensor (R_{conf}). The distance R_{conf} is correlated with the cubic root of the mass of gaseous mixture (M). Hence, the confined scaled distance is calculated by $Z_{conf} = R_{conf}.M^{-1/3}$.

To compare the overpressure measured in a confined environment with the corresponding one in free field at the same distance, the scaled distance Z_{conf} is used in Eq. (4).

5.1 Ground level

At ground level, for all the configurations with no opening (**Figure 13a**), the overpressure is always higher than the overpressure in free field (without confinement). Examination of the case with no opening means that the transmitter room is of course the only room studied. No amplification is obtained on gauge J in configurations 1, 2, and 4. The effect of the mobile wall close to gauge J in configuration 3 leads to a stronger overpressure. Configuration 2 leads to a ratio of 4 at position F, which is higher than the corresponding ratio in configuration 1. For gauges I and L, the ratio varies on average between 1.3 and 2.0.

The presence of an opening in the mobile wall means that two rooms need to be considered: a transmitter room and a receptor room (**Figure 13**).

In the receptor room, the small opening efficiently decreases the overpressure. A ratio of $\Delta P/\Delta P_{FF}$ below unity is reached on gauges I and L for configurations 2 and 3 and near 1 on gauge G for configuration 3. For configuration C4, the maximum overpressure is higher than the corresponding free field. Gauge G is directly exposed to the propagation of the shock wave, whereas gauge F is located at the limit of direct view. Hence, the overpressure on gauge G is higher than for gauge F. The mitigation decreases with the opening size. So, for gauge L the ratio varies from 0.55 to 0.7 for the small and large opening. The other positions in the receptor room show an overpressure higher than in free field, particularly on gauge G: the ratio on gauge G is 1.16 in C3 and 3.7 in C4.

In the transmitter room and configuration 2, the overpressure decreases on gauge F with the increase in the opening size. The large opening reduces the maximum overpressure on gauge G and the overpressure remains unchanged with the small opening compared to the same configuration with the full wall. The larger opening generates an increase in the overpressure on gauge L in configurations 4 and 3 and on gauge I in configuration 2.

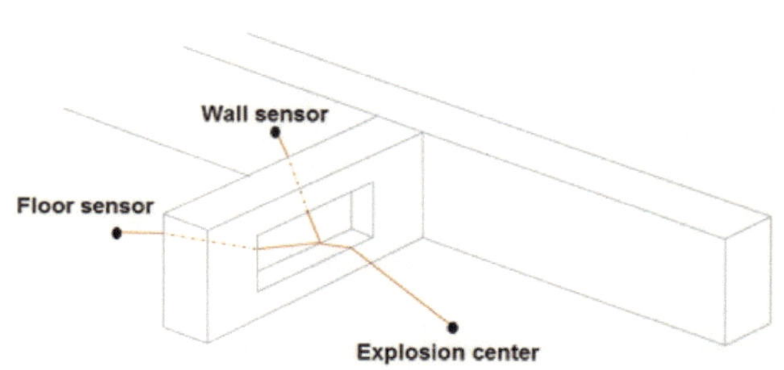

Figure 12.
Example of shorter path through the opening.

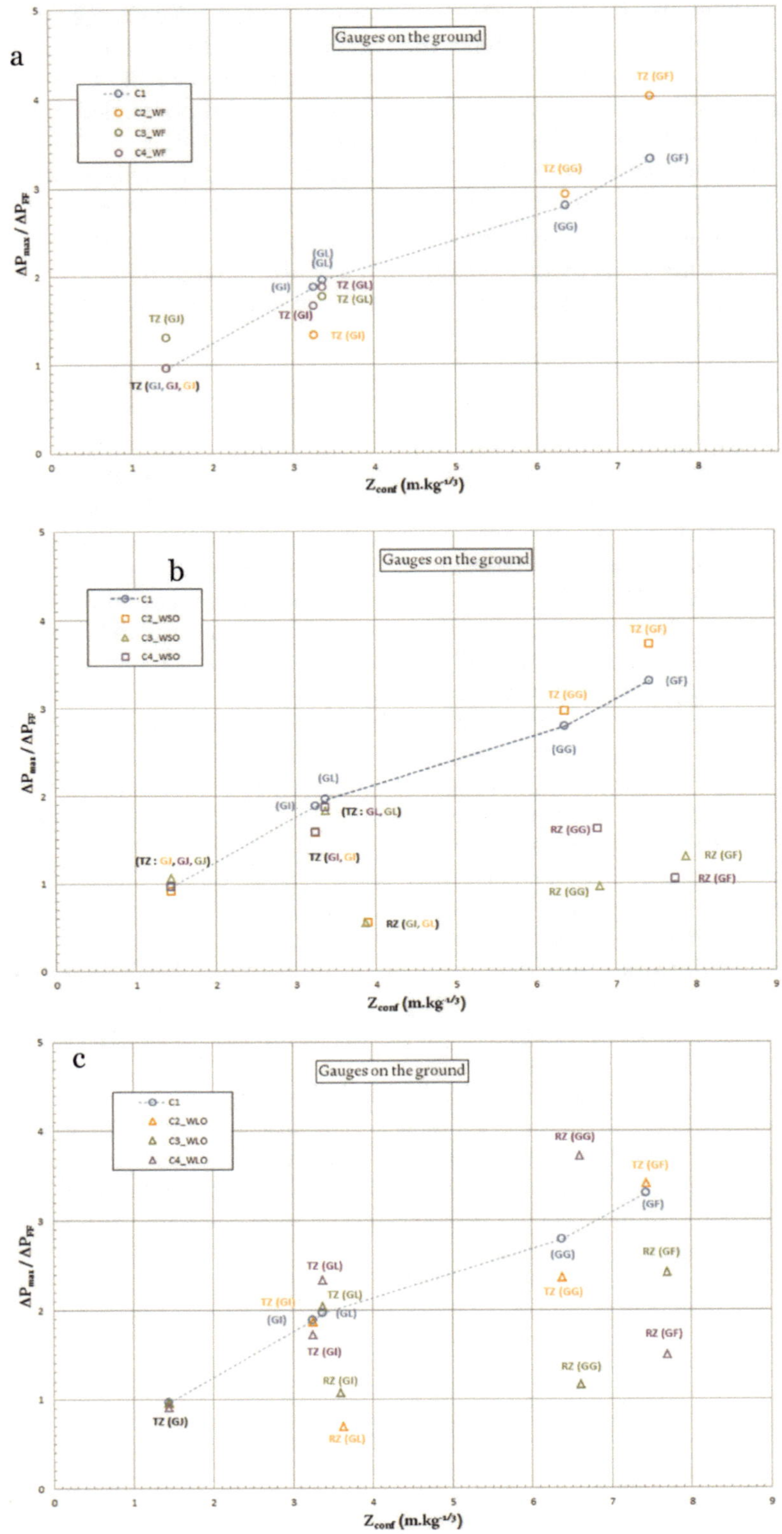

Figure 13.
Ratio of the maximum overpressure at ground level to the overpressure in free field versus the scaled distance in confined room. Pressure histories: (a) with full wall; (b) with small opening for C2, C3, and C4; and (c) with large opening for C2, C3, and C4.

5.2 On the walls

The overpressures were measured on the east wall with gauges located at half height of the wall. As expected, the overpressures are higher on the wall than on the ground. With the full wall (**Figure 14a**), gauge D is in the unprotected zone for configurations 1 and 2.

The ratio $\Delta P/\Delta P_{FF}$ is in the range 2 and 3 for all the other gauges and configurations. This level is higher than the overpressure level recorded on the ground for the same configurations and for a reduced distance $Z \leq 7\,\text{m kg}^{-1/3}$. Considering the smaller opening (**Figure 14b**), the gauges in the receptor room are protected, and the ratio of overpressure does not exceed 1.6 bar.

For gauges C and O, the ratio is lower than unity, which is consistent with the measurement recorded on the ground for the same configurations with, respectively, gauges I and L. The large opening (**Figure 14c**) does not mitigate the overpressure lower than in the free field, but it decreases the overpressure at location D with respect to configuration 1. Consequently, on the wall, the overpressure ratios are in the same order as on the ground. The large opening is efficient and limits the ratio to 3 with respect to the free field. Gauge D presents a dangerous location for configuration 1 as it is obtained with gauge F on the ground.

5.3 Effect zones on the ground

As defined in the French pyrotechnic legislation and the ministerial order No. DEVP0753277A on 20 April 2007 and DEVP0540371A on 29 September 2005 and the applicable version on 6 May 2019 [29], all explosives may result in five effect zones classified according to the damage caused to persons and property (**Table 2**).

Table 3 gives all the maximum overpressures obtained for all configurations and all gauges.

The table clearly highlights that the pressure field is mainly higher than 0.43 bar, which means that damage zone Z1 is dominant. Cases with $\Delta P_{max} < 0.43$ bar are indicated in green and correspond to the receptor zone with small opening in the mobile wall except for three cases in configuration 3 (gauge G) and configuration 4 (gauges F and D). Outside the zones isolated by a full movable wall, the maximum overpressure is never below 0.2 bar. This means that in these domains the damage zone is classified in Z2.

5.4 Applications to TNT charges and scale-up to large scale

The investigation proposed here can be applied to a TNT charge. The comparison of the pressure effect between the gaseous mixture ($C_3H_8 + 5O_2$) and the TNT charge is shown in **Figure 15**. The evolution for the TNT charge was extracted from the Unified Facilities Criteria [30]. In **Figure 15**, the incident experimental overpressure at ground level (gauges L, J, I, and G) is reported.

The comparison between the two explosive charges is expressed by the TNT equivalency, as detailed in [26]. This pressure-based concept is only considered for the object of this investigation. The equivalent mass of an explosive pressure is given by the mass ratio of TNT (M_{TNT}) to the considered explosive (M) that produces the same peak overpressure at the same radial distance of each load, hence:

$$E_{P\text{-}TNT} = \frac{M_{TNT}}{M} = \left(\frac{Z}{Z_{TNT}}\right)^3 \tag{5}$$

where Z is the scaled distance.

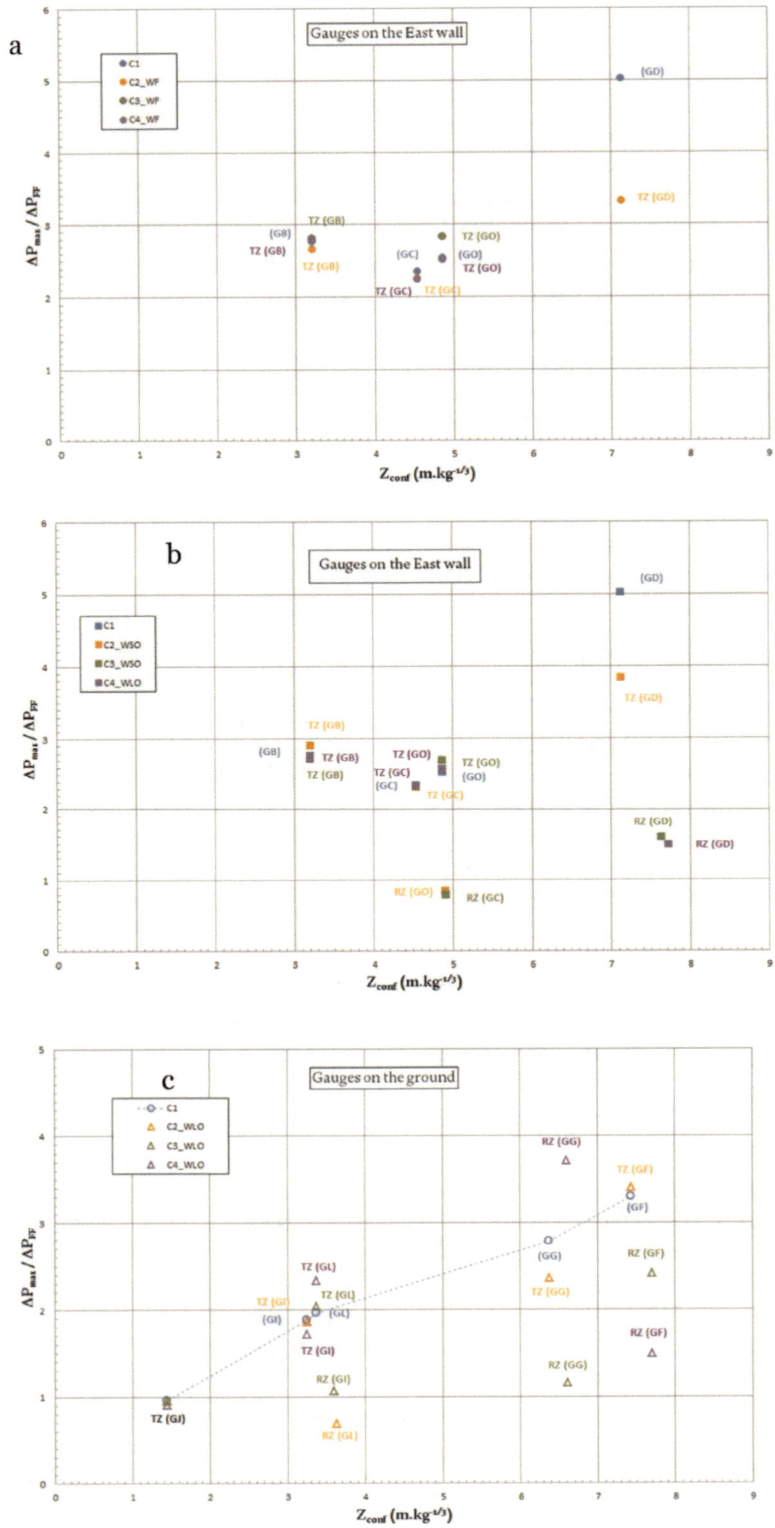

Figure 14.
Ratio of the maximum overpressure on the walls to the overpressure in free field versus the scaled distance in confined room. Pressure histories: (a) with full wall; (b) with small opening for C2, C3, and C4; and (c) with large opening for C2, C3, and C4.

Zone designation	Overpressure (bar)	Human casualties	Property damage
Z1	$\Delta P_{max} \geq 0.43$	Extreme (death 50%)	Extreme
Z2	$0.43 < \Delta P_{max} \leq 0.20$	Very serious	High domino effect
Z3	$0.20 < \Delta P_{max} \leq 0.14$	Serious	Serious
Z4	$0.20 < \Delta P_{max} \leq 0.14$	Significant	Light
Z5	$0.20 < \Delta P_{max} \leq 0.14$	Indirect injuries by breaking glass	Destruction of window

Table 2.
Effect zones for overpressure [29].

ΔP_{max} (bar)	C1	C2	C2-WSO	C2-WLO	C3	C3-WSO	C3-WLO	C4	C4-WSO	C4-WLO
Gauge L	1.6	/	**0.3 r**	0.5 r	1.4	1.5 t	1.6 t	1.5	1.5 t	1.9 t
Gauge J	3.5	3.5	3.3 t	3.4 t	4.7	3.8 t	2.5 t	3.5	3.5 t	3.3 t
Gauge I	1.6	1.1	1.3 t	1.6 t	/	**0.4 r**	0.8 r	1.4	1.3 t	1.5 t
Gauge G	0.8	0.8	0.8 t	0.7 t	/	**0.2 r**	**0.3 r**	/	**0.4 r**	1 r
Gauge F	0.7	0.8	0.8 t	0.8 t	/	**0.3 r**	0.5 r	/	**0.2 r**	**0.3 r**
Gauge O	1.1	/	**0.4 r**	0.6 r	1.2	1.2 t	1.1 t	1.1	1.1 t	1.1 t
Gauge B	2.4	2.3	2.5 t	2.3 t	2.4	2.4 t	2.0 t	2.4	2.4 t	2.4 t
Gauge C	1.1	1.1	1.1 t	1.1 t	/	0.4 r	0.6 r	1.1	1.1 t	1.1 t
Gauge D	1.2	0.8	0.9 t	0.7 t	/	**0.4 r**	0.5 r	/	**0.4 r**	**0.4 r**

Notations: r, receptor zone; t, transmitter zone; /, isolated zone; in bold, ΔP_{max} < 0.43 bar.

Table 3.
Maximum overpressure in bar for all configurations.

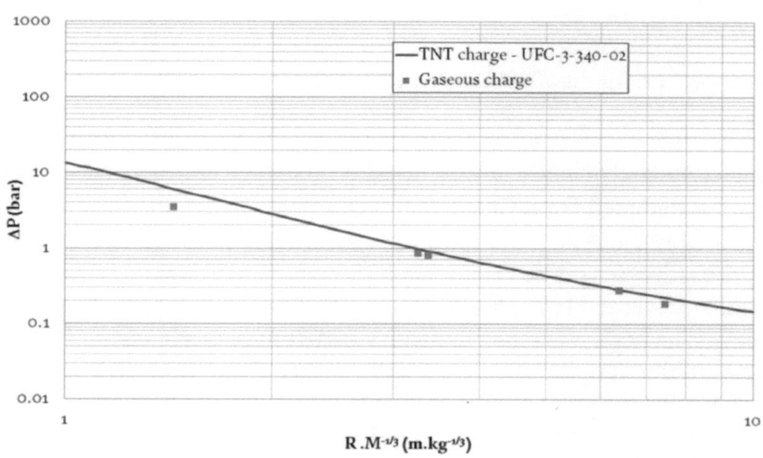

Figure 15.
Comparison of overpressure versus scaled distance for TNT and gaseous charge.

A TNT mass in the range of 0.49–0.96 g is deduced by applying Eq. (5).

The comparison of reflected overpressure was performed on the basis of abacus reporting in UFC-3-340-02 [30] and converted here into SI units (**Figure 16**).

For each gauge on the wall, the incident overpressure ΔP_i corresponding to the scaled radial distance is calculated from Eq. (4).

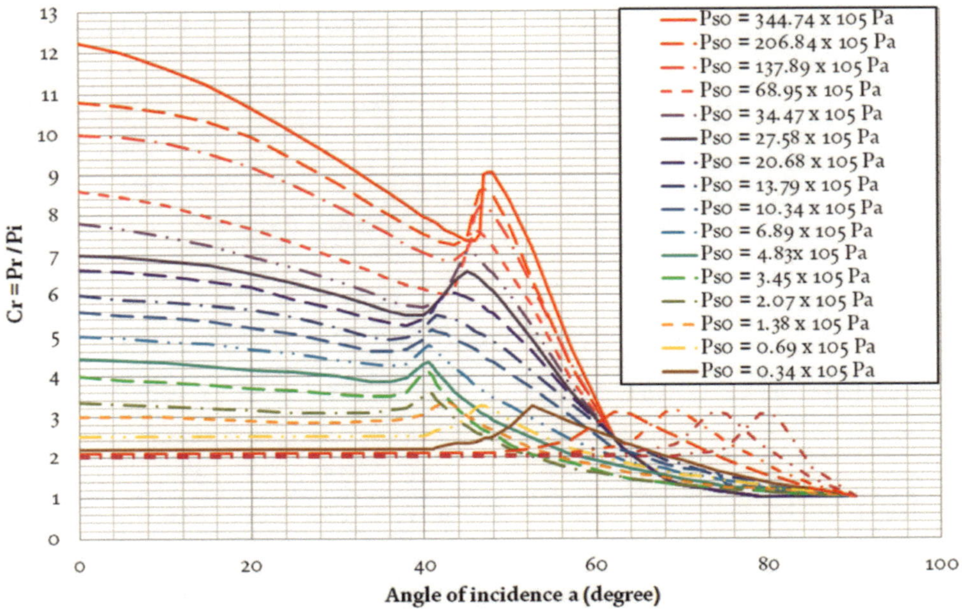

Figure 16.
Reflection coefficient angle of incidence and incident overpressure P_{so} (extracted from [30]).

Gauge	Z (m kg$^{1/3}$)	ΔP_i calculated (bar)	Cr abacus	ΔPr (bar)	ΔPr exp. (bar)	Deviation (%)
O	4.869	0.435	2.28	0.991	1.106	10.5
B	3.217	0.866	2.60	2.249	2.415	6.9
C	4.542	0.487	2.32	1.131	1.099	2.9
D	7.130	0.238	2.15	0.511	0.523	2.4

Table 4.
Comparison of reflected overpressures deduced from TNT abacus and measured.

The reflection coefficient is interpolated on the TNT abacus, and the reflected overpressure is deduced from $Cr = \frac{\Delta P_r}{\Delta P_i}$ and compared to the measurement. **Table 4** reports this analysis and demonstrates the validity of small-scale experiments. The maximum deviation is 10.5%.

The results presented here can be extrapolated by applying the Hopkinson law [31]. At scale 1, a shock wave results from the detonation of a mass of explosive M and impacts a point at a distance R from the center of the explosive charge. The shock wave arrives at a time Ta and is characterized by an overpressure ΔP^+, a positive and a negative impulse I^+ and I^-. The test is reproduced at scale k. Considering a mass of explosive kM, at a distance kR, the pressure profile presents the same overpressure ΔP, and the arrival time is multiplied by the scale factor k and hence the impulse kI^+ and kI^- too. Consequently, with this similitude law, the radial distance does not change with the scale:

$$Z = \frac{R}{M^{1/3}} = \frac{kR}{\left(k^3 M\right)^{1/3}} \tag{6}$$

The geometry of the target is also adapted by the scale factor k.

6. Conclusion

An experimental study of a pressure blast wave in a confined room facility was performed. The experiments were conducted at small scale with a gaseous charge. The facility simulated a single-story building with two rooms. The wall separating the two rooms was full or with an opening area. The effect of the location of this wall and the size of the opening were investigated. The pressure profile and the maximum overpressure were analyzed.

The pressure profiles are very complex to analyze. As expected, the overpressure increases with the confinement, and the protective effect of the wall decreases with the increasing size of the opening area. The most dangerous locations in terms of overpressure are at gauges J (ground level) and B (east wall) for all configurations due to their proximity to the explosive charge. In contrast, the least exposed zone corresponds to gauges G and F, which are the farthest from the charge but near a corner (for F) and near a wall (north) for G. Hence, the damage effects are severe since the maximum overpressure is never lower than 0.2 bar. The impulses were also examined, and values of 0.9 bar ms can be reached. The range of impulse level for the whole set of gauges and configurations is 0.3–0.6 bar ms. Severe damage results for high pressure and impulse.

Although this study is limited to a small scale and gaseous detonation charge, it is shown how the results can be applied at large scale and for a TNT charge. Numerical simulations would be interesting to complete this work.

Acknowledgements

The authors gratefully acknowledge the financial support of the Direction Générale de l'Armement Techniques Navales (DGATN) under the project 2016-058852 TN/SDT/PFN.

Nomenclature

CE	center of explosion (of explosive charge)
Ci (i = 1, 2, 3 or 4)	configuration i (1, 2, 3 or 4)
C_r	reflection coefficient
$E_{P\text{-}TNT}$	TNT equivalency in terms of pressure
h	height
h_M	height of Mach stem
HOB	height of burst
k	scale factor
L	length
Lw	width
M	mass of the gaseous mixture
P_0	ambient pressure
P_r	absolute reflected pressure
R	radial distance
R_{conf}	confined radial distance
R_i	distance from the wall
R_{i0}	transition distance of formation
RR	receptor room
TR	transmitter room
$V_{RR\text{-}Ci}$	volume of receptor room in configuration i

$V_{TR\text{-}Ci}$	volume of transmitter room in configuration i
WF	full wall
WLO	wall with largest opening
WSO	wall with smallest opening
Z	scaled distance
Z_{conf}	confined scaled distance
Z_i (i = 1,2,3,4 or 5)	zone designation
Greek	
Δ_{max}	maximum angle of oblique reflection
ΔP^+	positive overpressure
ΔP_{FF}	positive overpressure in free field
ΔP_i or P_{s0}	incident overpressure
ΔP_{max}	maximum overpressure

Author details

Isabelle Sochet[1*], Kevin Gault[1] and Luc Hakenholz[2]

1 INSA Centre Val de Loire, Université d'Orléans, Bourges, France

2 Direction Générale de l'Armement Techniques Navales, Toulon, France

*Address all correspondence to: isabelle.sochet@insa-cvl.fr

References

[1] Weerheijm J, Van Wees RMM, Absil LHJ, De Bruyn PCAM, Karelse JW. The fireworks disaster in Enschede: Overview, reconstruction, safety and pyrotechnics. In: Proceedings of European Safety and reliability Conference (ESREL); Maastricht, The Netherlands; 2003. Vol. 2

[2] Willye RJ. West fertilizer company fire and explosion: A summary of the U. S. chemical safety and hazard investigation board report. Journal of Loss Prevention in the Process Industries. 2017;**49**:132-138

[3] Zhao B. Facts and lessons related to the explosion accident in Tianjin Port, China. Journal of the international society for the prevention and mitigation of natural hazards. 2016;**84**: 707-713

[4] ARIA. Feedback on Technological Accidents [Internet]. 2019. Available from: https://www.aria.developpement-durable.gouv.fr/?lang=en

[5] U.S. Department of the Army. Fundamentals of protective design for conventional weapons, Technical Manual 5855. 1986

[6] U.S. Department of the Army. Structures to resist the effects of accidental explosions, Technical Manual 5-1300. 1990

[7] Kinney GF, Graham KJ. Explosive Shocks in Air. Berlin Heidelberg: Springer-Verlag; 1985

[8] Remennikov AM. A review of methods for predicting bomb blast effects on buildings. Journal of Battlefield Technology. 2003;**6**:5-10

[9] Edri I, Savir Z, Feldgun VR, Karinski YS, Yankelevsky DZ. On blast pressure analysis due to a partially confined explosion: I. experimental studies. International Journal of Protective Structures. 2011;**2**:1-20

[10] Gebbeken N, Döge T. Explosion protection architectural design, urban planning and landscape planning. International Journal of Protective Structures. 2010;**1**:1-21

[11] Benselama A, Mame J-PWL, Monnoyer F. A 1D–3D mixed method for the numerical simulation of blast waves in confined geometries. Journal of Computational Physics. 2009;**228**: 6796-6810

[12] Reichenbach H, Neuwald P, Kuhl AL. Role of precision laboratory experiments in the understanding of large-scale blast phenomena. In: 17th International Symposium on Military Aspects of Blast and Shock (MABS), Julius J. Meszarus Lecture; Las Vegas, Nevada, USA; 2002

[13] Baker WE. Explosions in Air. Austin, TX and London, UK: University of Texas Press; 1973

[14] Ohrt A, Rogers J, Oliver C. Investigation of internal air blast propagation at small scale. In: 24th International Symposium on Military Aspects of Blast and Shock (MABS), Halifax, Canada; 2016

[15] Ram O, Nof E, Sadot O. Dependence of the blast load penetrating into a structure on initial conditions and internal geometry. Experimental Thermal and Fluid Science. 2016;**78**: 65-74

[16] Togashi F, Baum JD, Mestreau E, Löhner R, Sunshine D. Numerical simulation of long-duration blast wave evolution in confined facilities. Shock Waves. 2010;**20**:409-424. DOI: 10.1007/s00193-010-0278-7

[17] Milne AM, Cargil SB, Longbottom AW. Modelling of complex blast. International Journal of Protective Structure. 2016;**73**:325-339. DOI: 10.1177/2041419616661431

[18] Massoni J, Biamino L, Jourdan G, Houas L, Igra O. Experimental and numerical investigation of blast wave interaction with a three level building. Journal of Fluids Engineering. 2017; **139**(11):111106-111106-9. DOI: 10.1115/1.4037172

[19] Miura A, Mastuo A, Mizukaki T, Shiraishi T, Utsunomiya G, Takayama K, et al. Reflection and diffraction phenomena of blast wave propagation in nuclear fuel cycle facility. JSME International Journal Series B: Fluids and Thermal Engineering. 2004;**47**: 287-292. DOI: 10.1299/jsmeb.47.287

[20] Jiang Z, Wang C, Miura Y, Takayama K. Three-dimensional propagation of the transmitted shock wave in a square cross-sectional chamber. Shock Waves. 2003;**13**: 103-111. DOI: 10.1007/s00193-003-0197-y

[21] Binggeli E, Binggeli F, Schlapfer D, Bucher KM. Airblast predictions in tunnel/entrance configurations due to he-detonations near the tunnel portal. In: 13th International Symposium on Military Applications of Blast Simulation (MABS), The Hague, The Netherlands; 1993

[22] Rigas F, Sklavounos S. Experimentally validated 3-D simulation of shock waves generated by dense explosives in confined complex geometries. Journal of Hazardous Materials. 2005;**A121**:23-30

[23] Sauvan PE, Sochet I, Trélat S. Analysis of reflected blast wave pressure profiles in a confined room. Shock Waves. 2013;**22**:253-264

[24] Julien B, Sochet I, Vaillant T. Explosion in a multi-chamber: Experimental investigation. In: 22nd International Symposium on Military Aspects of Blast and Shock (MABS) Bourges, France; 2012

[25] Julien B, Sochet I, Vaillant T. Impact of the volume of rooms on shock wave propagation within a multi-chamber system. Shock Waves. 2016;**26**:87-108

[26] Sochet I, editor. Blast Effects-Physical Properties of Shock Waves. Series: Shock Wave and High Pressure Phenomena. Springer International Publishing AG; 2018

[27] Vanderstraeten B, Lefebvre M, Berghmans J. A simple blast wave model for bursting spheres based on numerical simulation. Journal of Hazardous Materials. 1996;**46**(2–3):145-157

[28] Rose TA, Smith PD. Oblique clearing: Blast loads on buildings at non-zero angles of incidence. In: 18th International Symposium on Military Aspects of Blast and Shock (MABS); Bad Reichenhall, Germany; 2004

[29] French Pyrotechnics Legislation [Internet]. 2019. Available from: https://www.legifrance.gouv.fr/affichTexte.do?cidTexte=JORFTEXT000000245167&categorieLien=cid

[30] U.S. Army Corps of Engineers, Naval Facilities Engineering Command, Air Force Civil Engineer Support Agency. Technical Manuals, Unified Facilities Criteria (UFC), Structures to resist the effects of accidental explosions, UFC 3-340-02, 2008

[31] Hopkinson B. British ordnance board minutes 13565. 1915

Chapter 3

Direct Numerical Simulation of Nano Channel Flows at Low Reynolds Number

P. Srinivasa Rao

Abstract

The governing equations of viscous fluid flow are generally represented by Navier–Stokes (NS) equations. The output of Navier Stokes equations is in essence velocity vector from which rest of the flow parameters can be calculated. It is essentially a riotous task, sometimes it becomes so unmanageable that fluid flow over simplest topologies under low Reynold's numbers also needs the most powerful supercomputing facility to solve, if needed to model the fluid and its behavior under the turbulent conditions the best way out is to solve the averaged NS equations. However in the process of averaging Reynolds introduced certain new terms such as Reynolds Stresses. Therefore it is required to close the system of equations by relating the unknown variables with known ones. Hence we have turbulence models. Direct Numerical Simulation (DNS) is a method of solving NS equations directly that is by forfeiting the need of turbulence models as the equations are not averaged. However originally direct numerical simulation procedure does not need of additional closure equations, it is essential to have very fine grid elements and should be estimated for exceptionally small time steps to achieve precise solutions. In the present chapter an interesting flow through nano-channel problem has been discussed using the indispensable mathematical technique of computational fluid dynamics (CFD) which is DNS.

Keywords: direct numerical simulation (DNS), Navier stokes (NS) equation, turbulent flow, nano-channel flow Reynolds number

1. Introduction

While studying any physical phenomena be it of practical or academic importance, it is not always possible to mimic the exact system. For an example, if we want to study the flow of air past a moving train engine, we might have to build a wind tunnel along with a provision for a railway track for the engine to move in and out. We should have sensitive measuring instruments fixed at different places to measure various parameters. Again, finding the right locations for fixing these instruments can only be determined by carefully performing large number of iterations because the margin for errors in real-life problems are miniscule. For example, if the paint on the car is not done correctly even in some parts, it may contribute hugely to skin-friction drag and thereby reducing the mileage of the car.

One might argue that we can use scaled models keeping the physics of the problems same. These models induce errors during extrapolation to the actual situation besides being very costly in terms of infrastructure, resources and time. If we try to write a mathematical function and solve it, we should know all the pertinent parameters and laws that govern the flow. Then we might use various techniques like dimensional analysis and arrive at an equation. The validation of this mathematical equation again, can only be done by experiments. Once we get an equation, solving it for various cases is in itself a major challenge because we almost always end up with differential, integral or Integro-Differential equations. Till date, exact solutions to some of these equations are still open problems. So, approximate solutions are desired as against exact solutions. The advent of advanced computers brought along the solution for this in the form of Computational Fluid Dynamics (CFD). We can now analyze various types of fluid flows with numerical simulations and also develop suitable simulation algorithms. In most practical problems, the state variables of the fluids can be treated as continuous functions of space for which we already have conservation laws for mass, momentum, and energy. For many fluids that are used in engineering applications, there are well-accepted constitutive relations.

The fundamental fluid mechanics is put in the form of Navier Stokes equations way back in 1800s and thereupon many attempts were made to find the general solutions for the equations but even today it is an open millennium problem [1–3]. Based on the physics of interest, we make certain approximations to Navier–Stokes equations and therefore obtain the governing partial differential equations to be solved [4]. When tried to solve such governing equations using a numerical methods a suitable discretization procedure will be adopted and the problem will be attacked to solve using a better solver. Here we convert the mathematical model into a discrete system of algebraic equations. We even call it to be a difference equation in some gut words while the chosen method to solve the equations is Finite difference method (FDM). Many of this kind of solving methods have been evolved and to mention a few the finite element method (FEM), the finite volume method (FVM), boundary element method (BEM) are a few popular methods among so many numerical techniques. Out of many factors that influence selection, some are very crucial like problem defined on the geometry, the analyst's preference, and the predominant trend of solution may be in a selected and particular area of the problem space. Nonetheless it will be decided later about which numerical algorithm to be adopted and solve these equations and develop a computer programs [5–7]. After the advent of enormous speed in graphics processing units the computer graphics and animations formed from large output of numerical values make the visualization of the simulation results easier to human comprehension and draw the conclusions henceforth.

The nature of fluid flow is such a complicated phenomenon no matter what governing equations we formulate and try even more numerical solutions for the same, it can well be understood with a single classic statement of legendary Nobel Laureate R. P. Feynman in his popular scientific texts, in a quote "The efforts of a child trying to dam a small stream flowing in the street and his surprise at the strange way the water works its way out has its analog in our attempts over the years to understand the flow of fluids. We have tried to dam the water up—in our understanding—by getting the laws and the equations that describe the flow" depicts how novice the formulations are. The statement given by Feynman has already been proved more than quite often times in the applications of Fluid Dynamics in modern engineering and science, and more profoundly to the computational fluid dynamists.

> *Nonetheless an additional unique feature relates to the mathematical nature of Navier-Stokes equations. Steady-state Navier-stokes equations are elliptic in nature, whereas unsteady Navier-Stokes equations are parabolic in nature. As numerous problems associated with the methods of solution of the completely defined elliptic Partial differential equations the Navier-Stokes equations generally solved as an unsteady problem even in the case if the flow is steady, using a time marching schemes [8–10]. The transient solution of Navier-Stokes problem yields the solution but not be any easy compared to the steady state flow problem in more general geometric case.*

Turbulent flows involve randomly fluctuating flow variables such as velocity, pressure, and temperature.

Within the defined flow decomposition in accordance with Reynolds, a flow variable at a given spatial point at a given instant can be represented as the sum of a mean value and a random fluctuation about this mean value, and the process of obtaining the average or a mean value is represented as the technique of *Reynolds averaging*. Therefore, for any flow variable ϕ , its spatio-temporal variation can be expressed as stated below.

$$\phi\left(x_i ,t \right) = \overline{\phi\left(xi \right)} + \phi'\left(x,t \right)$$

Where $\overline{\phi}$ the mean or an averaged value and ϕ' is the variations in the property which is with fluctuating values or statistically steady turbulent flows.

$\phi(x)$ is the time average defined as

$$\overline{\phi\left(xi \right)} = \lim_{T\to\infty} \frac{1}{T} \int_0^T \phi(xi)dt$$

Where T represents the averaging time interval, which must be large compared to the typical time scales of fluctuations throughout the flow length.

For unsteady flows, ϕ represents ensemble averaging defined as

$$\phi\left(x_{i,} ,\mathrm{t} \right) = \lim_{N\to\infty} \frac{1}{N} \sum_{i=0}^{N} \phi(x_i ,t)$$

where N is the number of identical experiments.

The Reynolds averaging applied to the continuity and momentum equations of an incompressible fluid flow would be obtained in the following set of equations.

$$\frac{\partial(\rho\overline{vi})}{\partial xi} = 0$$

$$\frac{\partial(\rho\overline{vi})}{\partial xi} = \frac{\partial(\rho\overline{vi}\,\overline{vj})}{\partial xj} = -\frac{\partial\overline{p}}{\partial xi} + \frac{\partial\rho\left(\overline{\tau ij} - \tau ij\,R\right)}{\partial xj}$$

Where $\hat{o}_{ij}\mathrm{R} = -\tilde{n}\overline{\left(\acute{v}i\,v\acute{}j\ \right)}$ is Reynolds stress tensor.

The Reynolds averaged transport equation for a scalar ϕ

$$\frac{\partial(\rho\overline{\phi})}{\partial t}+\frac{\partial(\rho\overline{vj}\,\overline{\phi})}{\partial xj}=\frac{\partial\left(\Gamma\frac{\partial(\rho\overline{\phi})}{\partial t}+qjR\right)}{\partial xj}$$

Where $q_i R=\overline{\rho v' j\phi'}$ is turbulent flux.

The higher order correlations appear in Reynold's stress equation is due to non-linearity of Navier–Stokes equation, despite the fact that we try an attempt to derive the governing equation for higher-order terms, they would always result in equations with even higher order correlations.

The Eddy viscosity models are evolved basing on the Boussinesq proposition, which defines the Reynolds stresses which would be proportional to mean velocity gradients, as that of Deviatoric Reynolds stress. This stress is proportional to the mean rate of strain. Numerical computation of eddy viscosity possibly would be required for the solution of the set of transport equations.

Thus, the Reynolds averaged NS model turbulence models are going to be categorized on some of the several additional partial differential equations that are to be solved to enforce closure and find the final flow behavior in any of the defined topology. The zero-equation model is actually the one that computes the eddy viscosity solely from the Reynolds-averaged velocity field. The transport equation that is solved simultaneously with the Reynolds-averaged Navier–Stokes equations to determine the eddy viscosity is in other words known as one-equation model that refers to the family of turbulence models [11–13].

The highly established standard k -ε model which has been developed and contributed by Launder and Spalding makes use of two model equations, one for the turbulent kinetic energy k and the dissipation rate of turbulent kinetic energy per unit mass ε. The transport equations are solved using the Reynolds averaged Navier–Stokes equations. The k-€ turbulence model is well established and the most widely validated and uses the eddy Viscosity turbulence model [14]. Diverse modifications of the standard model have been recommended to account for the near-wall flow region and low Reynolds number turbulent flows. Nevertheless the Eddy viscosity models have significant deficiencies, which are due to some eddy viscosity assumptions. The experimental measurements and numerical simulations have indicated that in the cases of three dimensional turbulent flow the eddy viscosity turns out to be a tensor quantity [15]. Therefore, the use of a scalar eddy viscosity for computation of Reynolds stresses is might not be appropriate. It would as an alternative be recommendable to compute Reynolds stresses directly using its dynamic transport equations. This idea has been the basis of the Reynolds stress model though it the most expensive methods of the turbulence models in use in the present day CFD solutions and particularly for RANS simulations. It is relatively expensive computationally in comparison to eddy diffusivity models since it requires solving seven additional PDEs for every grid solution in this case. Hence these models are not used as widely as expected from the analysis's the eddy diffusivity models in practical analyses.

In the later days one of the most adoptable and practically sensible methods that have surfaced is Large Eddy simulation (LES). The method in which the Large scales of motion or large eddies of fluids which possess more energy than a smaller one have been taken to be most critical in the solving for turbulent flow cases and has been proved to be more effective in most of the cases both in the experimental and industrial applications [16]. The most effective transporters of conserved quantities viz. mass, momentum, and energy remain to be only large eddies compared to the smaller eddies. The momentum, and energy exchange is so minimal in such cases that they can be neglected. Moreover the behavior of these smaller scales of motion is universal in turbulent flows irrespective of the flow's context and geometry.

It is moderately undemanding to capture the effect of smaller eddies through a model. The underlying principle of LES is to treat the large eddies of the flow precisely and model the more universal small scale eddies, however this method is inherently time-dependent and are highly suitable for three-dimensional simulations. Such approach is less costly than DNS but a lot more expensive than RANS model for the same flow. Large Eddy Simulation (LES) is mostly preferred method for obtaining accurate time history for high Reynolds number and complex geometry flows. It separates the larger and smaller eddies through spatial filtering operation [17].

The conceptual steps are somewhat like this, it begins with the spatial filter to decompose velocity field $v_i(x,t)$ into a sum of resolved components $v_i(x,t)$ and residual sub grid scale component $v_i(x,t)$ $v_i(x,t)$ represents the motion of large eddies. Such equations are very similar to the original Navier–Stokes equation except for an additional residual stress term arising from filtering. Thus it will be obtained that the filtered velocity, pressure, and temperature field using an appropriate Navier–Stokes solver. In large eddy simulation, the spatial filtering operation for any transported field is defined by a using filter function. In LES, a filter kernels has been used, this also includes a Gaussian filter, a top-hat or box filter, and also spectral cut-off filter. Filtering operation roughly implies that eddies of a size larger than cut off width are large eddies while eddies of size smaller must be modeled. The convective term in the preceding equation cannot be computed in terms of the resolved velocity field. To get an approximation for this term, let us introduce the so called sub-grid stress tensor. The sub grid scale (SGS) Reynolds stress represents large scale momentum flux caused by small or unresolved scales, and it must be modeled to ensure closure. The dissipation takes place is much smaller than the characteristic length scale considering the length scale η of the flow with its ratio $\frac{L}{\eta}$ and this being proportional to $\mathrm{Re}^{3/4}$. As we allow the Reynolds number of the flow is increased length scales range it is observed the space and time in the flow and hence it becomes wider.

On the other hand due to the rapid growth in the computational resource Direct Numerical Simulation (DNS) of turbulent flows has become one of the crucial tools in turbulence research. Though DNS has gained its importance and has been universally recognized about its technique and methodology the kind of resources it consumes to figure out turbulence in some of the fundamental geometries is in itself a complicated affair.

It is also able to provide statistical information difficult to obtain by experimental measurements. Among effects to be observed in the DNS prior to laboratory experiments about flow velocity and vorticity vectors such effects are not present in if the modeling is done gaussian fields. However the attainment of high Reynolds number flow analysis necessitates the use of subgrid scale model to represent the effects of the unresolved small-scale turbulence on the explicitly designed and simulated large-scale flow. The leading length scales are considered to be of the order of higher dimension of flow region in the fluid flows or of the size of largest Eddie leading to turbulence in the fluid flow. The nominal scale of size of eddies in the inertia forces and viscous forces are of comparable magnitude so that the energy is cascaded to the Eddie, it is directly dissipated. Consecutively to simulate whole scales in turbulent flows, the computational domain must be sufficiently larger than the largest characteristic length scale of the flow L and the grid size must be smaller than the finest turbulence scale η.

The turbulent flows are essentially three-dimensional in nature and to satisfactorily model it needs at least require $(L/\eta)^3$ grid points, seems to be proportional to $Re^{9/4}$. The calculations must proceed for time scale of largest Eddie i.e., L/U, while the time steps must be of the order η/U which implies Re^n time steps must be taken per each run of the simulation. If we calculate the computing time necessary for the range would be of years for a single run in the limits of the magnitudes of Reynolds numbers encountered in practice. Most DNS performed does not take into account the vortices with size smaller than the grid size. To be able to obtain meaningful results from DNS sufficiently fine spatial resolution must be acquired to ensure that flow phenomena taking place at scales smaller than the grid resolution are negligible. For these reasons, DNS in many cases should be regarded as "highly accurate turbulent calculations without any use of turbulent models." The mathematical framework and numerical results therefore obtained must serve as the useful exact solutions of the physical problems only possible for limited cases and topologies for example in the region of laminar flow of fluid through a pipe and moreover that is possible if the nonlinear terms are dropped from the N-S equation which allows analytical solution in such confined and flow regions and geometries.

In the present work, turbulent flows have been selected to investigate using large eddy simulation (LES) principles and also to understand the direct numerical simulation methodology to model and study the turbulence in fluid flow in nano channels of an electronic device. The nano channel is designed to transit the fluid and therefore removes the heat from the plate. The fluid considered to be multiphase as there would be a base fluid in which the nanoparticles will be suspended and hence an Eulerian multiphase as well as VoF techniques is being implemented. The same conditions are been attempted to solve directly using the techniques of DNS code of Geris flow solver for a simple 2D model and an unified flow field has been created as an initial flow conditions to capture the turbulent flow along the length scale and adaptive mesh refinement has been adopted while selecting the flow solvers in the three dimensional flow generation in such nano channel. The problem uses a finite volume method (FVM), in which the conservation equations in their weak formulation are solved in a discrete number of cells, determining one value for the flow parameters in each cell. The visualization has been taken in paraview and the exported combined file is processed as a data file in commercial code to make a comparative study about the two well developed turbulence modeling techniques. The nano channel considered as a rectangular pipe flow shown in **Figure 1**. The turbulent channel flow is often referred to as a canonical flow of Poiseuille flow case, since it is one of the simplest flows one can think of to attack using the direct numerical simulation. Therefore, this test case is suitable to verify if a numerical solver is able to accurately predict the turbulent vortices near the wall of the tube particularly at the bottom of the channel as the fluid is incompressible.

Simulations between all the Reynolds numbers in the range of $100 \geq Re \geq 320$ the grids have been selected in the stream wise, the wall-normal, and the pipe length of the computational domain are $L_x \times L_y \times L_z = 2 \times 2 \times 40$ as shown in the **Figure 1**. The grid points have been $N_x \times N_y \times N_z = 178 \times 98 \times 356$. It has been proposed as the standard condition that the computational grid is uniform in the direction of stream flow. In due course of transient time approximately 18 convective time scales it has been observed that the steady state has been reached.
From this condition the time averaging has started and the flow has been simulated for more than 120 convective time steps statistically independent time levels have been considered for averaging flow velocities and vorticity in the length scale.

In the **Figures 2, 3** and **4** it has been shown about the convergence of the statistical data in the cases of turbulent intensity along the pipe length as the flow becomes steadily turbulent.

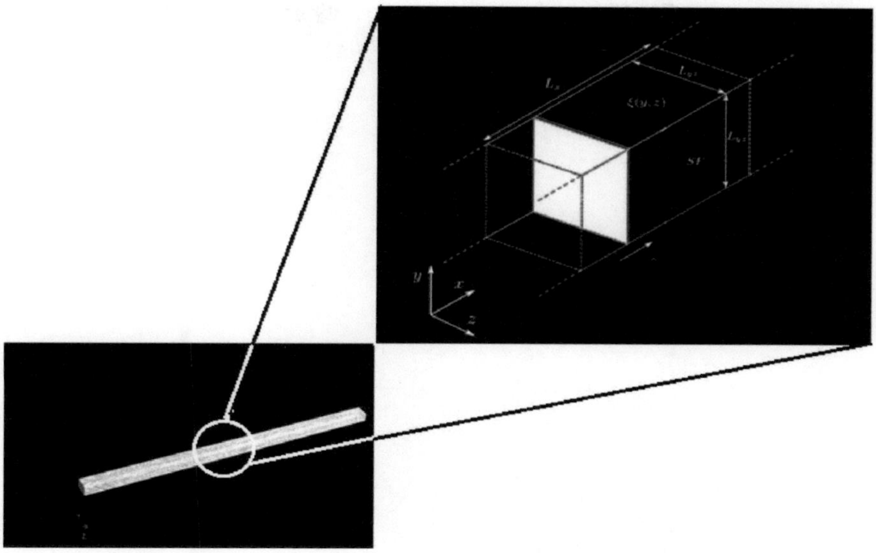

Figure 1.
Geometry of pipe.

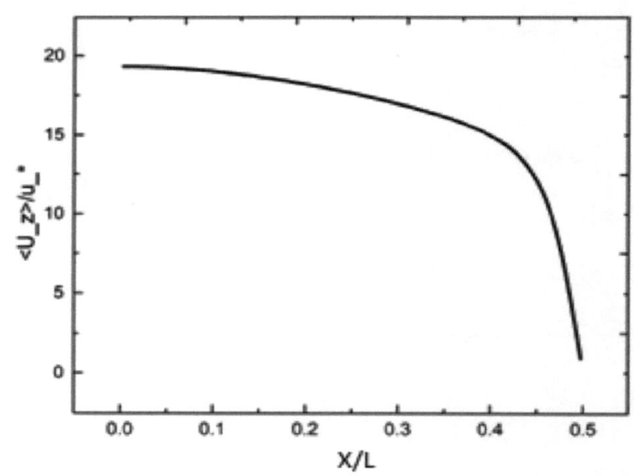

Figure 2.
Mean axial velocity.

The gradual development from uniform turbulence to a pattern at Reynolds number180 has been accounted and presented in the **Figure 4**. The energy distribution plays a key role in the flow transition and henceforth is to be tracked quantitatively and with the probability distribution. The velocity gradients near the pipe walls after the flow is fully developed would exhibit the axial flow velocity as shown in **Figure 2** and consecutively the root mean square velocity computed is presented in **Figure 3**. These plots are developed on the basis of non dimensional number along the length of the pipe. At the maximum fluid velocity along the channel it has been recorded that there is no significant raise or dip in the Reynolds number in all the runs.

The propagation velocity of the pattern is approximately that of the mean flux and is a decreasing function of Reynolds number. Examination of the time-averaged flow shows that a turbulent band is associated with two counter-rotating cells stacked in the cross-channel direction and that the turbulence is highly

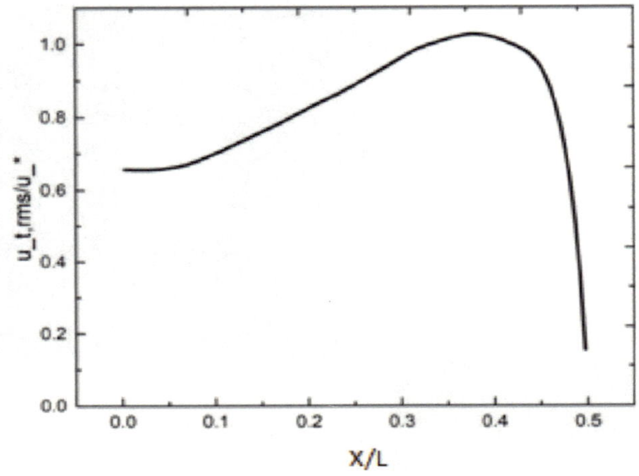

Figure 3.
RMS value along the pipe length.

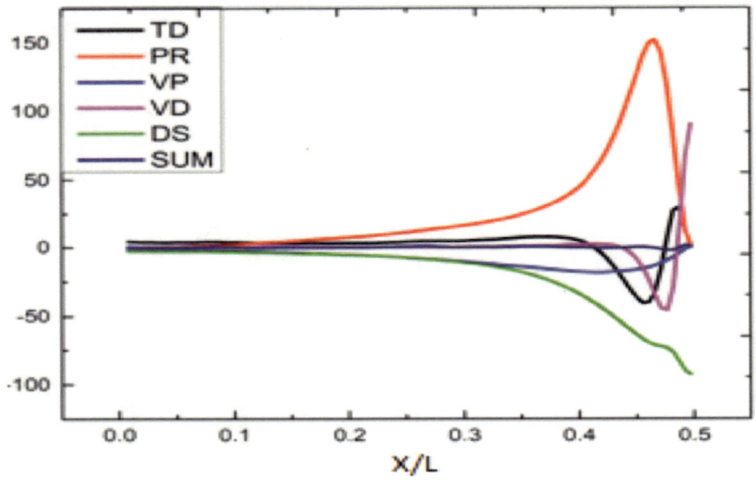

Figure 4.
Turbulent KE distribution record.

concentrated near the walls. Near the wall, the Reynolds stress force accelerates the fluid through a turbulent band while viscosity decelerates it; advection by the laminar profile acts in both directions. In the center, the Reynolds stress force decelerates the fluid through a turbulent band while advection by the laminar profile accelerates it. These characteristics are compared with those of turbulent-laminar banded patterns in plane Couette flow.

The average Reynolds number varies in the different cases, from 120 in the bare flow case to 512 in the fully developed convective flow conditions as a working nanofluid cooling device. Though the primary aspect of heat transfer is not considered until the turbulent flow did not notice during the flow through the channel, the swirling velocity component is completely ignored. The pressure drop is exactly getting equated to the experimentally verified case in the case of LES calculated value in almost all the cases as shown in **Figure 5**. Once the LES model could capture the significant turbulence properties, the problem is attempted on geris flow solver code for DNS solution for one such case to be verified with the already obtained results. Some of the results obtained are presented in **Figure 6**.

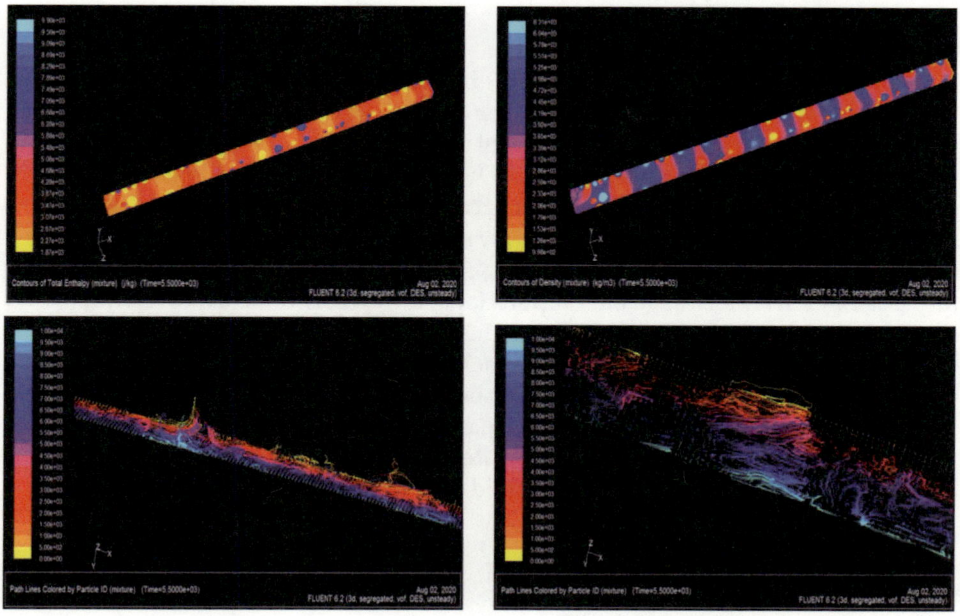

Figure 5.
LES based simulation.

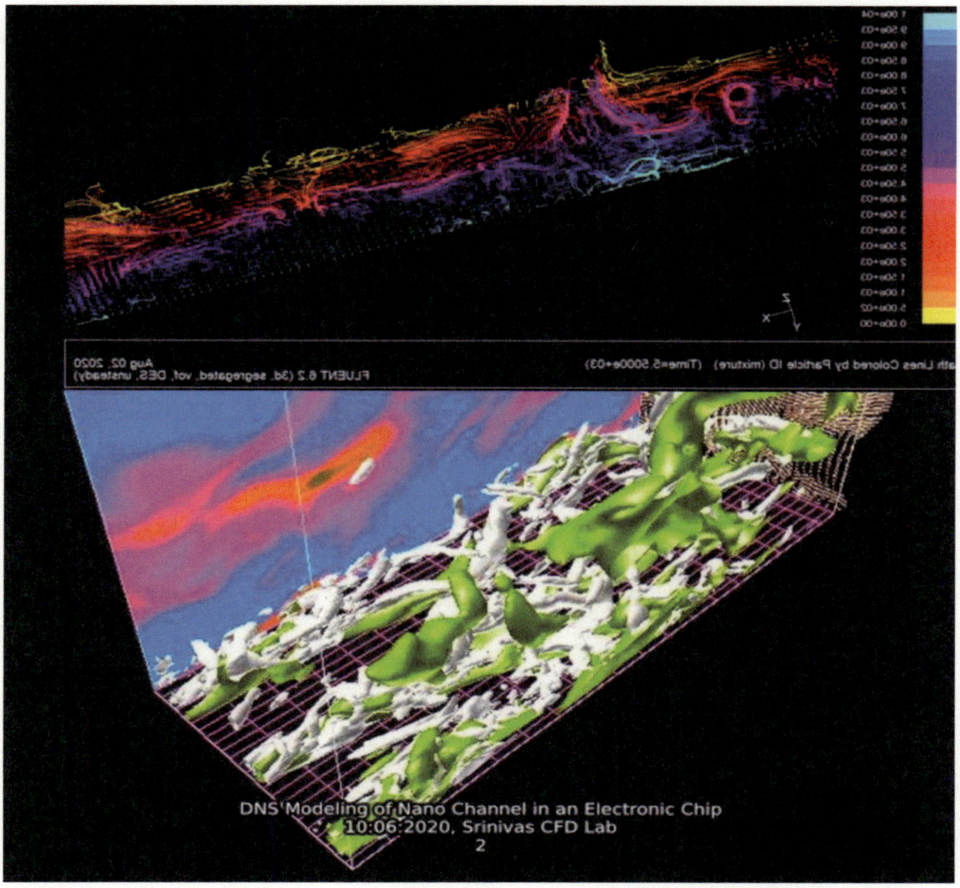

Figure 6.
DNS on LES based simulation.

2. Conclusion

A direct numerical simulations (DNS) has been performed over a nano channel considering as that of a nono pipe used for electronics cooling purpose with various set of Reynolds numbers and the effects of Reynolds number on the turbulence properties have been investigated. Since DNS of experimental-scale setups are beyond the author's computational resources, DNSs on a reduced computational domain were performed. The unified flow solver will be taking the job of adaptive mesh refinement (AMR) with automatic selection of fluid solver that has been implemented in whole range of computational domain. In contrast to the micro-channel flow the nano channel flow case has been much qualitatively trivial once the LES solution is available and further solving for a DNS solution of the problem at any particular Reynolds number. It has been shown that Kolmogrov length and velocity scales are more appropriate compared to the other such approximations while modeling DNS at significantly low Reynolds numbers including any other energy diffusion during the course of fluid flow through rectangular channels.

Author details

P. Srinivasa Rao
Department of Mechanical Engineering, Vardhaman College of Engineering, Hyderabad, India

*Address all correspondence to: professorpsrao@vardhaman.org

References

[1] Batchelor, G. K. 1953 The Theory of Homogeneous Turbulence. Cambridge University Press

[2] Orszag, Steven A. 1970 Analytical Theories of Turbulence. Journal of Fluid Mechanics. 41 (1970): 363-386.

[3] Garnier, E.; Adams, N.; Sagaut, P. (2009). Large eddy simulation for compressible flows. Springer. Publications

[4] Kraichnan, R. H. & Panda, R. 1988 Depression of nonlinearity in decaying isotropic turbulence. Phys. Fluids. 31, 2395.

[5] Leith, C. E., & Kraichnan, R. H. 1972 Predictability of turbulent flows. J.Atmos. Sci. 29, 1041-1058.

[6] S. B. Pope, Turbulent Flows (Cambridge University Press, Cambridge, 2000).

[7] E. F. Toro, Riemann Solvers and Numerical Methods for Fluid Dynamics (Springer, Berlin, 1999)

[8] Boersma, B.J.: Direct numerical simulation of turbulent pipe flow up to a Reynolds number of 61,000. J. Phys. 318, 042045 (2011)

[9] Chin, C., Ooi, A.S.H., Marusic, I., Blackburn, H.M.: The influence of pipe length on turbulence statistics computed from direct numerical simulation data. Phys. Fluids 22, 115107 (2010)

[10] Guala, M., Hommema, S.E., Adrian, R.J.: Large-scale and very-large-scale motions in turbulent pipe flow. J. Fluid Mech. 554, 521-542 (2006)

[11] Popinet S. Gerris: a tree-based adaptive solver for the incompressible Euler equations in complex geometries. J Comp Phys 2003;190: 572.

[12] V. V. Aristov, A. A. Frolova, S. A. Zabelok, R. R. Arslanbekov, V. I. Kolobov - Simulations of Pressure driven flows through Channels and Pipes with unified flow solver

[13] Wosnik, M., Castillo, L., George, W.K., 2000.A theory for turbulent pipe and channel flows. J. Fluid Mech. 421, 115.

[14] Colin S., Rarefaction and compressibility effects on steady and transient gas flows in microchannels. Microfluid Nanofluid 2005; 1:268-79

[15] Sharipov F., Numerical simulation of rarefied gas flow through a thin orifice. J Fluid Mech 2004: 518:35-60

[16] Welty, J. R.; Wicks, C. E.; Wilson, R. E.; Rorrer, G. L., Differential Equations of Fluid Flow. In Fundamentals of Momentum, Heat, and Mass Transfer, John Wiley and Sons, Inc.: , 2008; pp 99-112.

[17] Srinivasa Rao P. (January 1st 2010). Modeling of Turbulent Flows and Boundary Layer, Computational Fluid Dynamics, Hyoung Woo Oh, IntechOpen, DOI: 10.5772/7108.

Section 2

Reactive Fluid Turbulence

Chapter 4

Bases of Combustion Instability

V.I. Biryukov

Abstract

Combustible systems generally consist of two types of chemically interacting components during combustion: an oxidizing agent (oxygen, fluorine, chlorine, their compounds) and fuel (hydrogen, hydrocarbons, nitrogen and hydrogen compounds, aluminum, etc.). The chemical properties of the components, their phase state, and their physical structure are essential when choosing the methods for supplying the components and organizing the processes in the combustion chambers, but they relatively weakly affect the basic laws of combustion processes. In the theory of combustion, the problems of burning homogeneous, premixed, gaseous components are studied in most detail. The concepts and methods of the theory of combustion are used in other areas of science and technology when considering exothermic processes with high heat generation. The separation of the issues of flame stability into diffusion-thermal and hydrodynamic problems, which is often encountered in theoretical works, is conditional and is caused by the desire to reduce the mathematical difficulties that arise when solving the problem in the general formulation. In fact, flame instability is determined by the influence of both transport processes in the flame (diffusion-thermal processes), depending on its structure, and hydrodynamic processes, i.e., the effects of gas flow. The determination of the concentration limits of flame propagation, ignition, and extinction, spontaneous instability of the flame front, the transition of combustion to detonation, and the excitation of oscillations during combustion are practical problems of the theory of combustion. Acoustic combustion instability can be considered as a self-oscillating process in which the feedback providing the energy necessary for maintaining undamped wave motions from a nonperiodic heat source (combustion process) is realized through the action of sound (acoustic) waves on combustion; in this case, the parameters of the wave motions, amplitude, waveform, and frequency, are determined by the internal properties of the system itself. This chapter provides a sequence of parametric estimates of acoustic instability during combustion in cylindrical chambers.

Keywords: combustion instability, intracameral instability, acoustic self-oscillations, modes

1. Introduction

Chemical transformations during combustion are accompanied by a strong increase in the temperature of the reacting substances, since the initial mixture has a large supply of chemical energy. These transformations usually occur in a thin (less than 1 mm), intensely luminous combustion zone, propagating at a certain speed (of the order of 10–10^3 cm/s) through the initial system. The combustion zone (flame) is a wave of exothermic chemical reaction—a combustion wave [1].

The most important feature of combustion processes is the strong exponential dependence of the rate of a chemical reaction on temperature. Because of this, at a usual initial temperature, the reaction rate in the initial system is negligible, and the system is in a stable, quasi-equilibrium state (although its reacted components correspond to its state of chemical equilibrium). With a significant increase in temperature, an exothermic reaction occurs, which can spread through the medium. As a result of a large heat release, the nearest layers of the combustible mixture are heated, initiating a quick reaction in them, and so a self-propagating combustion process occurs. Moreover, in the combustion zone that separates the cold initial mixture from the hot combustion products, large temperature and concentration gradients are formed, causing heat transfer through heat conduction and diffusion of the initial mixture and combustion products. Thus, the development of an exothermic chemical reaction during combustion is accompanied by physical processes of heat and substance transfer, which, in turn, have a strong influence on the intensity of the reaction and on the speed of flame propagation. Combustion in power plants and engines is usually carried out in a moving medium, and this affects the nature of the movement. In turn, the movement of the medium affects the processes of heat and substance transfer and, consequently, combustion. The influence of physical factors is especially pronounced when the fuel components are separately supplied to the combustion chamber—fuel and oxidizer, which is widely used in engines. In this case, the combustion process is dependent on mixture formation—evaporation (in the case of liquid components) and mixing of the components, as a result of which the combustion characteristics in many cases are determined mainly by physical processes. If one of the components of the fuel during combustion is in the condensed phase, then there may be heterogeneous combustion in which a chemical reaction occurs at the interface. Then the combustion characteristics are determined mainly by heat exchange processes, the rate of supply of reacting substances to the interface between molecular and convective diffusion. Thus, although chemical transformation is the basis of combustion, the physical processes of heat transfer, transfer of reacting substances, and gas-dynamic conditions play an important role in it. The physical interpretation of the combustion processes makes it possible to summarize the extensive experimental data and observations, establish the general laws of combustion, and obtain practical results. The concepts and methods of the theory of combustion are also used in various fields of science and technology when considering exothermic processes with high heat release. The phase state and physical structure of the starting components can be different: it is possible to burn gaseous components, burn a liquid atomized component in a gaseous stream of another component, burn liquid components and combustion of solid components, etc. The chemical properties of the components, their phase state, and physical structure are essential in selecting methods for feed components and organizing process in the combustion chamber but relatively little effect on the basic laws of combustion processes [2–6].

2. Chemical combustion reactions and relaxation processes in the combustion zone

In the combustion processes that have to be dealt with in technology, characterized by a large increase in temperature due to the release of chemical energy of the starting components in the form of heat, the combustion mechanism is thermal in nature, determined by the accumulation of heat in the reacting system. An increase in temperature in this case leads to an increase in the rate of each of the stages of the reaction, which increases the rate of the total reaction and the rate of heat release.

The resulting intermediate reaction products—atoms and radicals—play the role of active centers in accordance with the kinetic mechanism of the reaction; their effect on combustion cannot be separated from the effect of increasing temperature. In the case of complex combustible systems, widely used in power plants and engines, the chemical reaction mechanism includes a large number of stages and possible intermediate products. Clarification of the detailed kinetic scheme of the total chemical process during the combustion of complex systems is not always possible. Therefore, in most cases, one has to confine oneself to the simplest concepts of the kinetics of chemical combustion reactions. The speed of a homogeneous chemical reaction can be represented as a change in the concentration of the initial gas mixture per unit time:

$$W = -\frac{dC}{dt}$$

The rate of simple exothermic reactions considered in the theory of combustion depends only on temperature and on the concentration of the starting materials [4, 5]:

$$W = kC^n \exp\left(-\frac{E}{RT}\right),$$

where k is a constant coefficient (preexponential factor), C is the concentration of the initial gas mixture at a given moment, n is an indicator depending on the order of the reaction (for a first-order reaction, $n = 1$; second-order $n = 2$), E is the activation energy, T is the gas temperature, and R is the gas constant.

The main influence on the course of combustion processes and on the conditions for the transition of combustion from one level to another (ignition, extinction) is exerted by the ratio E/R. When burning gases, according to experimental data, E/R $\sim 15{\cdot}10^3$ K, and the temperature dependence of the rate of a chemical reaction is very significant. The exponential dependence of the rate of a chemical reaction on temperature is nonlinear, and this determines significant differences in the implementation of combustion. Irreversible processes occurring in the combustion zone—physical and chemical—contribute to the transition of the initial quasi-equilibrium system to the equilibrium state, and they can be considered as relaxation processes, and the characteristic transition time corresponding to each of them can be considered as relaxation times (transition processes to the equilibrium state are called relaxation). The relaxation time gives the time scale during which equilibrium is established for the process under consideration. The structure of the combustion zone depends on the relaxation processes occurring in it. The relaxation time can be defined as the residence time of the substance in the corresponding part of the combustion zone. In the simplest cases of burning mixed gases, the width of the combustion zone (in order of magnitude) is equal to $\delta = a/u$, and the residence time of the substance in this zone is $\tau = \delta/u = a/u^2$ [4]. Here a is the thermal diffusivity of the gas, and u is the flame propagation velocity. The value of τ determines the relaxation time to establish equilibrium in chemical composition during the combustion of perfectly mixed gases.

The study of processes in the combustion zone is facilitated by the fact that often they are very different from each other in relaxation time. This allows us to introduce simplified combustion models in which some state parameters under certain conditions are considered unchanged (frozen), while other parameters (or the same parameters but in other conditions) are considered to be of equilibrium value. As an example, consider a detonation wave propagating in a gas mixture (e.g., in a mixture of hydrogen with air), which initially has room temperature. The relaxation time of the chemical reaction in the case under consideration (i.e., the time of its transition to an equilibrium state in chemical composition into combustion products) is so long that the gas parameters in front of the wave can be considered unchanged. When an intense shock wave passes through the mixture, the temperature of the gas in the wave rises in a short time (of the order of the time of several collisions between molecules). Therefore, a shock wave can be considered as a discontinuity in which only an increase in temperature occurs with a constant chemical composition of the mixture. Behind the shock wave that heats the combustible mixture and high temperature, there is a fast chemical edition leading to the establishment of chemical equilibrium. In this case, the relaxation time is significantly larger than the shock compression time (of the order of 10^4 molecular collisions), and a relaxation layer forms behind the shock wave. Diffusion and thermal conductivity; as relatively slow processes in the relaxation zone can be neglected. Such is the model of the detonation wave developed by Zeldovich [4]. The analysis of this model instead of the extremely simplified model of the detonation wave as a gas-dynamic discontinuity made it possible to significantly develop the theory of detonation. Using as a parameter, the relaxation time of processes in various regions of the combustion zone is especially effective in the analysis of unsteady multistage combustion of complex systems, for example, solid rocket fuels. The comparison of the relaxation time of processes in the condensed phase with the relaxation time of processes in the gas phase with a rapid change in external conditions, for example, the pressure in the combustion chamber, makes it possible to draw important qualitative conclusions about the occurrence of unsteady combustion for a wide range of technical systems with a chemical transformation mechanism that has not yet been studied.

3. Critical combustion conditions

The main features of combustion include the presence of critical phenomena—sudden changes in the mode of the process with a small change in temperature, pressure, etc. The conditions under which this abrupt regime change occurs are called critical conditions. The consideration of the theory of thermal ignition proposed by Semenov [5] makes it possible to quantify the critical conditions of ignition. From the condition of equal heat release and heat removal at the ignition boundary, we can obtain the ratio for determining the temperature of the reacting body (heating from chemical reactions):

$$T_1 - T_0 = \frac{RT_0^2}{E}.$$

It is enough that the temperature of the gas mixture rises due to a chemical reaction by several percent, and a transition to a quick chemical reaction occurs from an almost complete absence of a chemical reaction. Depending on the conditions under which the reaction proceeds, qualitatively different realizations can be obtained. The accumulation of heat in the reacting system causes a sharp transition from a slow, almost imperceptible reaction to a progressively accelerating reaction. Such sharp transitions are in fact natural for simple exothermic reactions. They are associated with the exponential dependence of the reaction rate on temperature according to the Arrhenius equation, with a powerful effect on the reaction process of "feedback"—the accumulation of heat in the reacting system, increasing the temperature of this system. The obtained critical ignition conditions separate the region in which the quasistationary state of the system under consideration is possible from the region of the essentially unsteady process of burnout ("thermal explosion"). These critical conditions depend on the relationship between the heat release from the chemical reaction and the heat removal to the external environment, i.e., from the interaction of chemical and physical processes, which determine the impossibility of thermal equilibrium in the volume of a reacting combustible mixture with desired properties.

4. Volumetric combustion in a gas mixture stream

Combustion in a moving mixture usually occurs in a combustion wave propagating through the mixture and is characterized by the presence of inhomogeneities and transport processes. But, if such an intensive mixing of the initial mixture with combustion products is ensured in the flowing combustion chamber so that the parameters of the mixture are the same at all points of the chamber, then combustion in such a complete mixing chamber will proceed according to the volumetric mechanism.

Consider the possible stationary modes of the process in the chamber of complete mixing. Note that in the case of thermal ignition, the critical ignition conditions corresponded to the transition from the region of a quasistationary process to the region of nonstationary process of progressive burnout—"thermal explosion." In the case of a full mixing chamber, critical conditions determine the transition from one stationary level to another stationary level, and the duration of the existence of stationary states of the system is not fundamentally unlimited. We assume that at the entrance to the chamber, there is a gas with a concentration of C_0 and a temperature of T_a. In the chamber, due to complete mixing, the temperature T and gas concentration C remain unchanged; the same temperature and composition have a gas exiting the combustion chamber. Next, c is the relative concentration of the reacting substance (the mass of a given substance per unit mass of the mixture), which is associated with the volume concentration C by the ratio

$$c = \frac{C}{\rho},$$

where ρ is the density of the substance.

Loss of heat through the walls into the environment will not be taken into account.

To find possible stationary modes of the process in the combustion chamber, we study the equation of its heat balance. This equation, given that all the heat released as a result of a chemical reaction, goes to increase the thermal part of the enthalpy of combustion products (neglecting the kinetic energy of the latter due to its smallness), we write in the form

$$V_k WH = \dot{m} c_p (T - T_0), \quad (1)$$

where H is the heat of combustion of the mixture, V_k is the volume of the combustion chamber, W is the reaction rate, m is the second mass of gas passing through the combustion chamber, and c_p is the heat capacity of the gas (we consider it constant).

Chemical reaction rate $W = k(c\rho)^n exp\left(-\frac{E}{RT}\right)$. For simplicity, we assume that the reaction is first order ($n = 1$). Then the reaction rate, expressed through the dimensionless concentration, can be represented as

$$W' = \frac{W}{\rho} = kc\exp\left(-\frac{E}{RT}\right).$$

From the conditions of constancy of the total enthalpy, i.e., the sum of the thermal part of the enthalpy and chemical energy (combustion in the chamber proceeds at constant pressure and without heat exchange with the environment), we find the relationship between concentration and temperature in the form $C = c_0\frac{T_g - T}{T_g - T_0}$. Thus, as the temperature rises, the concentration of the reactant decreases (burnout).

Now the reaction rate equation can be written as

$$W' = k_1(T_g - T)\exp\left(-\frac{E}{RT}\right),$$

where $k_1 = \frac{C_0}{T_g - T_0}k$.

With a large value of the thermal effect of the reaction and, correspondingly, a larger increase in the temperature of the reacting gas, the prevailing effect on the reaction rate is temperature; a decrease in the concentration is much weaker. Therefore, the curve in the case under consideration has an exponential character and reaches a maximum only after 80–90% of the starting material is consumed.

These features of the temperature dependence of the reaction rate are the cause of sharp changes in the combustion mode. With low heat release, the effect of a decrease in concentration on the reaction rate is more significant than the effect of temperature. In this case, the transition from one combustion mode to another can occur gradually, smoothly. We rewrite Eq. (1) in the form

$$HW' = \frac{\dot{m}\cdot c_p(T - T_0)}{V_k\rho}, \qquad (2)$$

where HW ‘is the rate of heat release of the chemical reaction; $\frac{\dot{m}\cdot c_p(T-T_0)}{V_k\rho}$- the rate of heat removal from the combustion chamber by the flow of matter.

Possible solutions to the heat balance equation, where $q_{xp} = HW'$ curvilinear dependence and.

$q = \frac{\dot{m}c_p(T-T_0)}{V_k\rho}$- straight lines allow you to set stationary temperature levels of combustion due to various conditions. One can find out the effect on the combustion regime of changes in various parameters: m, V_k, T_0, H, etc. To do this, one should consider how the combustion regime will change in the case of a high thermal effect of the reaction with changing parameters.

$\frac{\dot{m}}{V_k\rho}$ and other conditions unchanged. It is easy to see that the expression $\frac{\dot{m}}{V_k\rho} = \frac{1}{\tau_0}$ is an important parameter of combustion analysis. Here τ_0 is the gas residence time in the combustion chamber.

5. Unsteady burning

Unsteady combustion processes include ignition, extinction, flame propagation limits, flame front instability (here we mean intrinsic, spontaneous instability), the transition of combustion to detonation, and oscillations during combustion. The study of these phenomena is of great practical interest. New tasks and problems in the field of combustion processes that have arisen as a result of the rapid development of aircraft engines, their thrust growth and, accordingly, their burning power, and the use of high-energy components relate mainly to issues of unsteady combustion, especially combustion oscillations. Unfortunately, the theory of unsteady combustion is not fully developed.

6. Flame front instability

The instability of the laminar flame front can be caused by changes in the flame structure due to the difference in the diffusion coefficients of the components of the combustible mixture, leading to a partial change in the composition of the mixture (enrichment of the mixture with a faster diffusing component) immediately before the flame front. Stratification of the mixture under the influence of selective diffusion causes local changes (increase and decrease) in the burning rate, as a result of which the smooth surface of the flame is bent, the flame breaks up into separate cells—a cellular structure of the flame is formed, and polyhedral flames appear on the burners. The flame stability is also influenced by the ratio between the diffusion coefficients D and the thermal diffusivity of the component in the mixture in deficiency: if $D > a$, and then from physical considerations, we can expect an increase in the flame velocity in areas that are convex towards the initial mixture and decrease in concave ones, which leads to instability flame, if $D < a$ flame will be stable. The instability of the flame depends not only on the structure of the flame and the diffusion-thermal processes occurring in it but also on the interaction between the flame and the gas flow in the immediate vicinity of it, i.e., from hydrodynamic processes [6, 7].

7. Unsteady phenomena in combustion chambers

In the combustion chambers of power plants, in rocket and aircraft engines, and in other plants with nonequilibrium processes accompanied by intense heat generation (nuclear reactors, chemical reactors with exothermic processes, etc.), there are always random pressure fluctuations with different frequencies, the amplitude of which can to achieve significant values (in rocket engines up to 3% of the average pressure in the combustion chamber of the Russian Federation). This kind of unsteadiness in combustion chambers—sound noise—is usually associated with hydrodynamic causes of the same nature as in ordinary turbulent gas flows. In fact, the nature of sound noise and the mechanism of its generation in combustion chambers are almost not studied.

In forced combustion chambers with high pressure values, in some cases, regular pressure fluctuations occur with a frequency close to the natural (acoustic) frequency of gas oscillations in the reaction volume and with an amplitude increasing in time (which can stabilize at some level). This type of unsteadiness—acoustic combustion instability—is associated with the excitation and amplification of acoustic (sound) waves in the combustion chamber. The interaction of acoustic waves with the combustion process causes fluctuations in the burning rate and, accordingly, fluctuations in the rate of heat generation, which under the conditions determined by the Rayleigh criterion [2, 3, 7–10] leads to the conversion of the heat of combustion into the mechanical energy of acoustic waves. The physically clear interpretation of acoustic combustion instability as a self-oscillating process with acoustic feedback is applicable to a wide range of problems of excitation and amplification of sound waves in heat-generating systems, in various types of combustion chambers. Sound noise in the combustion chambers is also a self-oscillating process in which the energy source is the combustion process and the feedback is through the influence of sound waves on combustion. In this case, undamped oscillations of a stochastic nature arise, having a wide frequency band and random phases. Excitation and amplification of sound waves in combustion chambers as a result of their interaction with combustion processes lead to the generation of sound noise and to acoustic combustion instability, accompanied by pressure fluctuations. The boundary conditions for wave disturbances are determined by the geometry of the path and the parameters of the regime. For example, the ratio of the length and diameter of the cylindrical combustion chamber, the configuration of the nozzle on one side and the nozzle head on the other, determine the conditions for the excitation of a certain mode of acoustic vibrations at some fixed frequencies corresponding to a certain vibration form (tone). The number of natural frequencies of the camera (path) is unlimited. The most essential for practice are the lower harmonics. In the study of their forms and frequencies of combustion chambers, its analogy with acoustic resonators is often used. However, the combustion processes of two-phase flows are neglected, or they are significantly simplified, the mixture formation provided by the specific nozzle head and the processes of the flow of combustion products in the subsonic part of de Laval nozzle are not taken into account. All this distinguishes wave phenomena in combustion chambers from processes in a pipe closed at one end. At the same time, the simplified task allows one to attract the well-known acoustics ratios for a closed cylindrical pipe with rigid walls [8–11], also the classic *Fundamentals of Acoustics* by E. Skuchik and many other works. In the framework of the linear theory, one can quite accurately take into account the influence of the geometry of the chamber and nozzle on the shape and frequency of natural acoustic vibrations. To evaluate the effect on the oscillatory process of the spatial distribution of combustion, as well as to obtain analytical relationships when acoustic absorbers are installed in the chamber, the amplitude of acoustic pulsations can theoretically be determined only on the basis of nonlinear theory. Linear approaches are convenient for analyzing the conditions for the excitation of oscillations and determining relations for the instability boundary. At the same time, nonlinear models describe the processes of growth of disturbances and transition to the state of limit cycles.

At present, there is no theory that takes into account all aspects of the combustion instability problem in engines; methods for calculating combustion stability at the engine design stage have not been developed. At the same time, the results of a theoretical analysis of simplified models of combustion instability in engines are successfully used in practical studies to predict the effect of changes in design and operating parameters on instability, to search for means to increase stability, and to conduct targeted experimental work when developing engines with respect to stability.

In the analysis of a number of problems associated with combustion instability in engines, a simplified description of the dynamics of the combustion process is successfully applied by introducing the delay time between the moment the fuel enters the combustion chamber and the moment it is converted into final combustion products. The model of the delay time is based on the replacement of the real burnout curve by a step function characterized by a single quantitative indicator—the delay time or relaxation time of the combustion process. With this approach, the complex and almost unstudied mechanism of unsteady combustion in engines is excluded from consideration. For the first time, the model of delay time was proposed by Natanzon in 1949 in solving the problem of low-frequency instability in short-threaded rocket engines [9]. Crocco and Chen [10] developed a model of pressure-dependent delay time (pressure-sensitive delay time model) and applied to the analysis of low-frequency and high-frequency (acoustic) combustion instabilities in liquid-propellant rocket (LPR). When assessing the stability of combustion in energy engines, the influx and loss of acoustic energy in the combustion chamber are determined. In the case of solid fuel engines, the determination of the influx of acoustic energy is simplified, since the combustion zone can be considered combined with the burning surface of the charge. The developed experimental methods using special model chambers (T-shaped chambers) make it possible to determine the acoustic conductivity of the burning surface for a given fuel and then calculate the influx of acoustic energy, taking into account the distribution of the amplitude of the oscillations for the considered acoustic mode over the burning surface [11]. The loss of acoustic energy in the combustion chamber can be approximately calculated from known acoustic relationships and determined experimentally on a model of the combustion chamber.

8. Acoustic combustion instability

Intracameral self-oscillations can be modeled by a system of two dynamic links: a combustion link (a system of equations describing the conversion processes) and an acoustic link closed by positive feedback. The input coordinate of the combustion unit is the pressure perturbation δp and the gas flow velocity perturbation δu and the output perturbation of the flow rate of the combustion products δG. The dimensionless perturbations of the parameters appear to be related to the stationary mean values, for example [12]:

$$\delta p = \frac{p'}{p_0}; \delta\rho = \frac{1}{\gamma}\delta p; \delta u = \frac{u'}{u_0}. \quad (3)$$

Here γ is the adiabatic exponent, $\delta\rho$ is the perturbations of gas density, and δu is the pulsations of the gas flow rate.

Flow perturbations are associated with perturbations of the density and gas flow rate, such as

$$\delta G = \delta\rho + \delta u. \quad (4)$$

The perturbations arising in the combustion zone of the axial components of the mass flow density lead to additional pressure pulsations that propagate to the nozzle and are reflected from the surface of its subsonic part. Thus, the dynamic properties of the acoustic link are characterized by its frequency response θ, which has the form

$$\delta p = \theta(\delta G(t)). \quad (5)$$

The combustion unit can be represented by a nonlinear dependence

$$\delta G(t) = H(\delta p(t)), \quad (6)$$

where H is the frequency response of the combustion unit.

For high-frequency acoustic vibrations arising in the considered dynamic system, which constructively constructs a cylindrical combustion chamber with a short subsonic part of the nozzle, the values of the vibration frequencies correspond to the resonant maxima of the frequency characteristic of the acoustic unit. In turn, the vibration modes (modes) are close to the vibration modes of a conservative system similar to the link under consideration (**Table 1**).

Fashion fluctuations	m	n	ν_{mn}	Fashion fluctuations	m	n	ν_{mn}
First tangential	1	0	1.841	First radial	0	1	3.883
Second tangential	2	0	3.054	Second radial	0	2	7.16
Third tangential	3	0	4.2	3-я Радиальная	0	3	10.174
Combined	1	1	5.332	Combined	2	1	6.708
Combined	3	1	8.015	Combined	1	2	8.527
Combined	1	3	11.707	Combined	2	2	9.97
Combined	2	3	13.171	Combined	3	2	11.346
Combined	3	3	14.588				

Table 1.
The values of the roots ν_{mn} for the case of transverse vibrations in a cylindrical combustion chamber.

Following the works [2] also the classic *Fundamentals of Acoustics* by E. Skuchik and many other works, we present the solution of the wave equation in cylindrical coordinates (x, r, φ) in the following form:

$$\delta p = \sum_{mnk} |\delta p|_{mnk} J_m \left(\nu_{mn} \frac{r}{R}\right) \cos m\varphi \cos\left(\omega_{mnk} t + \psi_{mnk}\right), \tag{7}$$

where $\sum_{mnk} |\delta p|_{mnk}$ is the amplitude of the mode (m, n, k) in the transverse and axial directions,

ω_{mnk} is the frequency of the vibration mode (m, n, k), ψ_{mnk} is the phase shift of the modes (m, n, k), R is the radius of the combustion chamber, $J_m(z)\left(z = \nu_{mn} \frac{r}{R}\right)$ is a Bessel function of the first kind satisfying the Bessel equation, and ν_{mn} are the roots of the equation $\left(\frac{d}{dz} J_m\right)_{z=\nu mn} = 0, (m = 0, 1, 2 \dots)$.

The acoustic link is almost linear [12]; therefore, the relationship of the input and output signal for it is determined by the ratio

$$\delta\overline{p}_{mnk} = \overline{\theta}(i\omega_{mnk})\delta\overline{G}_{mnk}, \tag{8}$$

where θ is the acoustic frequency response of the camera for the mode (m, n, k).

In a linear consideration of the instability problem of a given dynamic system, i.e., completing the harmonic linearization of formulas (6) and (7), we obtain the relation for the frequency response of the combustion unit

$$\overline{H}\left(i\omega_{mnk}, |\delta\overline{p}_{mnk}|\right) = \frac{\delta\overline{G}_{mnk}}{\delta\overline{p}_{mnk}}, \tag{9}$$

where the pressure perturbations have the form

$$\delta p = |\delta p|_{mnk} e^{i\omega_{mnk} t} J_m(\nu_{mn} r) \cos m\varphi \cos\left(\omega_{mnk} t + \psi_{mnk}\right), \tag{10}$$

or the same in complex form

$$\delta p = \delta\overline{p}_{mnk} e^{i\omega_{mnk} t} J_m(\nu_{mn} r) \cos m\varphi, \tag{11}$$

for each individual mode, the oscillation amplitude is $\delta\overline{p}_{mnk} = \frac{1}{2} |\delta p|_{mnk} e^{i\psi_{mnk}}$.

In this case, it is possible to analyze the dynamic properties of the combustion zone for each individual vibration mode, independently of the others.

The condition for the excitation of self-oscillations in the combustion chamber is represented by the complex equation:

$$\overline{H} \cdot \overline{\theta} = 1. \tag{12}$$

The frequency characteristics of the acoustic link and the combustion zone are uniquely associated with conductivity $\alpha = \frac{\delta \overline{u}}{\delta \overline{p}}$.

$$\begin{aligned} \overline{H} &= \frac{(1 + \alpha_{\text{гор}})}{\gamma}; \\ \frac{1}{\overline{\overline{\theta}}} &= \frac{(1 + \alpha_{\text{ак.з}})}{\gamma}. \end{aligned} \tag{13}$$

Here $\alpha_{\text{гор}} = \alpha^{+}_{\text{гор}}(\omega) + i\alpha^{-}_{\text{гор}}(\omega)$ is the conductivity of the combustion unit and.
$\alpha_{\text{ак.з}} = \alpha^{+}_{\text{ак.з}}(\omega) + i\alpha^{-}_{\text{ак.з}}(\omega)$is the conductivity of the acoustic link.

Eqs. (12) and (13) imply the equality for the real and imaginary parts of the conductivity of the system links

$$\begin{cases} \alpha^{+}_{\text{гор}}(\omega) = \alpha^{+}_{\text{ак.з}}(\omega), \\ \alpha^{-}_{\text{гор}}(\omega) = \alpha^{-}_{\text{ак.з}}(\omega). \end{cases} \tag{14}$$

Particular solutions of Eq. (11) corresponding to fixed values of the indices (m, n, k) correspond to different vibration modes. Longitudinal vibrations (m = 0; n = 0; $k \neq 0$) propagate along the chamber in the axial direction. In this case, the tapering part and the critical section of the nozzle are considered as boundaries from which pressure waves are reflected. The geometric characteristics of a combustion chamber with a supersonic nozzle include the length, diameter of the cylindrical part, the shape and dimensions of the subsonic part of the nozzle, and the shape and design of the nozzle head. The geometry of the combustion chamber determines the values of the natural frequencies of self-oscillations and the height of the resonance maxima of the amplitude-phase frequency response of the acoustic link. At the resonant frequency, the imaginary part of the conductivity for the acoustic link is equal to zero $\alpha^{-}_{\text{ак.з}}(\omega)\approx 0$, and the real one is characterized by the ratio $\alpha^{+}_{\text{ак.з}}(\omega) = \alpha^{+}_{\text{сопла}}(\omega) > 0$. It follows that the flow of acoustic energy at the exit of the acoustic link is equal to the flux emitted through the nozzle. The maximum value of the modulus of the frequency response of the acoustic link is achieved at frequencies equal to the values of the intrinsic resonances:

$$|\overline{\theta}|_{\max} = \frac{\gamma}{1 + \alpha^{+}_{\text{ак.з}}(\omega_{\text{p}})}. \tag{15}$$

The higher the conductivity of the nozzle, the lower the amplitude maximum of the frequency response of the oscillatory system. Changing the shape of the tapering part of the chamber leads to a change in the dispersion of the energy of acoustic vibrations, because energy losses are determined by the shape of the oscillating gas volume. The influence of the shape of the subsonic part of the nozzle on the chamber acoustics was studied in detail in [8–10, 12]. In the case of the implementation of transverse oscillation modes, the indices of Eqs. (7) and (8) take the following values: for radial (m = 0; $n \neq 0$; k = 0) and for tangential oscillations ($m \neq 0$; n = 0; k = 0) (**Figure 1**).

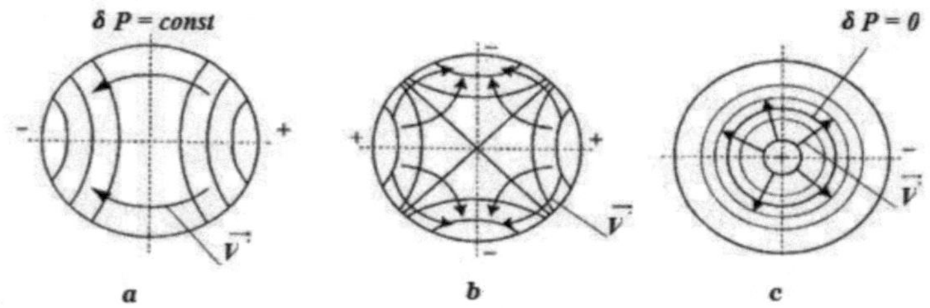

Figure 1.
(a) is a diagram of the propagation of velocity perturbations in the direction perpendicular to the axis of the combustion chamber, and at the same time one nodal diameter is depicted—this example characterizes the first tangential mode; (b) is a diagram of the propagation of velocity perturbations in the direction perpendicular to the axis of the combustion chamber, and at the same time two nodal diameters are shown—this example characterizes the second tangential mode; (c) is where the nodal circles are shown, but there are no nodal diameters. This example characterizes the first radial mode of acoustic vibrations in a cylindrical chamber.

In longitudinal plane waves, particle paths are parallel to the direction of flow. When transverse modes of HF vibrations are excited, the particles of the medium oscillate along trajectories perpendicular to the direction of gas flow. Standing waves can be formed by the interaction of direct and reflected waves having the same frequency. Aggregations in a standing wave are characterized by a cross section with a maximum value of the amplitude of pressure oscillations and a zero amplitude of velocity oscillations, and the nodes of the waves are points with a zero pressure amplitude and a maximum velocity amplitude. The distance between successive nodes (or antinodes) is equal to half the wavelength. Radial vibrations are axisymmetric and directed from the axis to the side wall of the chamber. The pressure antinode points in the radial mode of oscillation are located in the center of the chamber and on its walls. Therefore, when radial acoustic vibrations are excited, the displacement of the combustion products leads to an increase in the gas pressure in the center of the combustion chamber and a decrease near its walls. After a time equal to half the oscillation period, the pressure in the center of the chamber, where it was high, becomes low and back against the wall. For tangential modes, two waveforms are distinguished. One of them is standing, having a wave form that is invariable in space but with a varying amplitude. Another form is a traveling wave in which the entire mass of gas, including a line of nodes, rotates. In the cylindrical part of relatively short combustion chambers, it is most often the tangential modes of acoustic vibrations that are excited. Of the two tangential modes, running or standing, in a chamber without partitions, a running mode is more common. The consequence of their excitement is, as a rule, significant damage to the material part. A sharp increase in the rate of heat transfer is associated with transverse instability modes, which often causes burnout of the firing base, nozzle nozzles, or walls of the combustion chamber for a very short period of time, for example, 100–150 ms.

Heat flows during unstable combustion can increase up to 10 or more times in comparison with a stable combustion mode. The fight against acoustic instability is one of the main problems in the development of high-power aircraft and rocket engines, including during boosting the regime. Modern experimental methods for assessing stability in engines are based on the creation of artificial disturbances in the combustion chambers (using an explosive sample inside the chamber or an explosion product stream outside the combustion chamber, etc.) with subsequent registration and analysis of the resulting oscillatory processes [2, 3, 13]. The creation of artificial perturbations of pressure or flow (equal to speed) in the combustion chambers, and the subsequent response of the combustion processes recorded during the experiment, as well as the registration of natural "noise" and the subsequent analysis of the results of the statistical processing of the measured signals allow us to obtain quantitative estimates of the stability margins of the working process compared with other known approaches [3, 13]. The first method is similar to that used in mathematics to study the stability of systems of equations: an artificial perturbation is introduced into a working combustion chamber, in the other case, natural perturbations of the working process—"noise during combustion"—are used. For this, appropriate criteria were developed that quantitatively characterize the stability margins, which guarantee the engine's operability under all the required modes and external conditions stipulated by the terms of reference.

The analysis of natural "noise" provides a large amount of information about the stability margins of the working process and, in addition, allows for the serial production of engines to control production stability in terms of maintaining the stability margin.

Imagine the response of the combustion process in the chamber to an artificially created pressure impulse in it in the form of a sum of expressions [2]:

$$y = Ay_0 \exp\left(-\int_0^t \delta dt\right) \cdot \sin\left(\int_0^t \omega dt + \varphi_0\right), \tag{16}$$

where:

Ay_0 is the initial amplitude of the deviation.

$\delta = \delta\ (t)$ and $\omega = \omega\ (t)$ are the attenuation coefficient and oscillation frequency.

φ_0 is the initial phase of the oscillations.

The damping coefficient of oscillations has the following physical meaning:

$$\delta = (E_2 - E_1)/2E_{\Sigma.}, \tag{17}$$

where $E2$ is the flow of vibrational energy (per unit time) dissipated by the system, $E1$ is the flow of vibrational energy entering the system, and E_Σ is the total vibrational energy in the system.

It is convenient to replace the attenuation coefficient by a dimensionless quantity obtained by multiplying by a period, i.e., decrement of oscillations δT. Oscillations experimentally obtained in the combustion chambers of engines can be analyzed by presenting them in the form (16). The values of δ and ω fully characterize the oscillations; this allows us to use them as experimental indicators of stability.

To determine the experimental values of the decrement of pressure fluctuations, including noise, the following methods can be applied: spectral, correlation, amplitude, and the instantaneous method [13].

The identification of the spectrum of the measured vibration frequencies is carried out in order to isolate the natural frequencies of the acoustic path under study, on which vibration amplification (resonance) is realized during the development of the product. The main method is to compare the experimental frequencies with the calculated values for the natural modes of oscillation of the gas volume of the tract. Information on the values of various other parameters is necessary for statistics and clarification of their limit values, as well as, if necessary, changes to the design of the combustion chamber affecting stability.

To create algorithms for the quantitative assessment of the stability margin of the combustion chambers of rocket engines, a number of tasks were required:

- Selection of optimal indicators of process stability.
- Development of equipment for registration and processing of source information.

- Study of the relationship of stability indicators with structural and operational parameters of engines.
- Determination of the required values of indicators to ensure sufficient reserves of stability of the working process of engines of various schemes. The margin of stability with respect to the hard excitation of self-oscillations in the combustion chambers is sufficient if, after the introduction of a pulse perturbation, the pressure fluctuations decay quickly enough, and a certain relationship between the initial peak and the gas pressure in the chamber before its creation is satisfied. However, the selection of the optimal ratios of the parameters of the working process necessary to assess the sufficient stability margins with respect to the acoustic vibration modes required the processing of a large amount of statistical data for a number of engines [3].

The minimum impulse from artificial disturbance is selected taking into account the fact that it:

- Must exceed a value below which probabilistic excitation of the instability of the working process may occur.
- Should significantly exceed the level of ripple "noise" to ensure the calculation of stability parameters with sufficient accuracy. The maximum value of the initial pressure pulse should be limited due to the minimum interference in the working process of the chamber under study.

It was established that in order to assess the stability of the working process to finite disturbances, it is necessary [2, 3] to introduce a pressure pulse within $15\ A_{sv} < A_m < 25\ A_{sv}$ and determine the relaxation time t_p.

where:

A_m is the average value of the absolute maximum of the pulse.

A_{sv} is the average rectified value of the pulsations of natural noise.

$t_p = t_1 + t_e$ is the total relaxation time of the working process.

$t1$ is the time of exposure to the working process of the first wave of perturbation.

t_e is the time of decreasing the amplitude of the pressure fluctuations e times.

The combustion chamber is considered resistant to finite disturbances if $t_p < 15$ ms.

When testing full-scale engines, it is necessary to evaluate the decrements and spectra of pressure oscillations before the disturbance is introduced and after the oscillations are damped. The decrements of pressure fluctuations and spectra should not differ within the accuracy of measurements. Their significant difference will mean the instability of the work process.

Author details

V.I. Biryukov
Moscow Aviation Institute, National Research University, Moscow, Russia

*Address all correspondence to: aviatex@mail.ru

References

[1] Bulat PV, Upyrev VV. Detonation and its initiation—History of experimental, theoretical, and numerical research. Engineering-from theory to practice. In: A collection of articles based on materials of the XLVIII-XLIX International Scientific and Practical Conference No. 7–8 (44). Novosibirsk: SibAK. 2015. pp. 1-16. Available from: http://sibac.info

[2] Biryukov VI, Mosolov SV. The Dynamics of the Gas Paths of Liquid Rocket Engines. Moscow: MAI Publishing House; 2016. 168p

[3] Shibanov AA, Pikalov VP, Saidov SS. In: Denisova KP, editor. Methods of Physical Modeling of High-Frequency Instability of the Working Process in Liquid Rocket Engines. Moscow: Mechanical Engineering/Mechanical Engineering-Flight; 2013. 512p

[4] Zeldovich JB. Selected Works. Chemical Physics and Hydrodynamics. Moscow: Nauka; 1984. 374p

[5] Semenov NN. Chemical Kinetics and Chain Reactions. Moscow: Nauka; 1966. p. 604

[6] Raushenbah BV. Vibration Combustion. Moscow: GIFFL; 1961. 500p

[7] Markstein DG, Genosh G, Putnam AA. Unsteady Flame Propagation. Moscow: Mir; 1968. 437p

[8] Harrje DT, Reardon FH, editors. Liquid Propellant Rocket Combustion Instability. Washington: NASA; 1972. 870p

[9] Natanzon MS. The Instability of Combustion. Moscow; 1986. 274p

[10] Crocco L, Chen SI. Theory of Combustion Instability in Liquid Propellant Rocket Motors AGARSDograph #8. London: Butterworths Ship Pub; 1956. 351p

[11] Abugov DI, Bobilev VM. Theory and calculation of solid propellant rocket engines. Moscow; 1987. 272p

[12] Biryukov VI, Mosolov SI. Acoustics of gas paths of liquid rocket engines. Moscow: MAI Publishing House; 2013. 164p

[13] Biryukov VI, Tsarapkin RA. Experimental determination of the damping decrement in the combustion chamber of liquid rocket engines. In: Russian Engineering Research. Vol. 39. Allerton Press, Inc.; 2019. pp. 6-12

Chapter 5

Determination of the Velocity of the Detonation Wave and the Conditions for the Appearance of Spherical Detonation during the Interaction of Hydrogen with Oxygen

Myron Polatayko

Abstract

The well-known formula for the flat detonation wave velocity derived from the Hugoniot system of equations faces difficulties, if being applied to a spherical reactor. A similar formula has been obtained in the framework of the theory of explosion in reacting gas media with the use of a special model describing the transition of an explosive wave in the detonation. The derived formula is very simple, being also more suitable for studying the limiting processes of volume detonation. The conditions for the transition of a shock wave to a detonation wave are studied. Initial detonation conditions required for fast chemical reactions to take place at the front of a spherical explosive wave have been determined. A simple relation describing the critical detonation temperature for various pressures in the hydrogen-oxygen mixture was obtained. Using the known formulas for a shock transition, the critical temperature was coupled with the initial conditions in a static environment, such as the pressure, temperature, and hydrogen content in the mixture.

Keywords: detonation, point blast, spherical wave, Haber scheme, Lewis scheme, kinetics of chemical reactions, critical temperature

1. Introduction

The strong explosion in a small volume of a detonating gas mixture has been studied well in modern physics [1, 2]. The velocity of a detonation wave propagating in a spherical reactor can be calculated absolutely precisely with the use of a variety of original software programs [3]. There are also approximation formulas such as

$$\frac{D}{D_n} = 1 - \frac{A}{r - R_x}, \tag{1}$$

where r is the current radius, R_x is the critical radius, A is a constant, D_n is the velocity of a plane wave, and D is the velocity of a spherical wave. For larger charges, if the radius exceeds the critical one, the Eyring dependence

$$\frac{D}{D_n} = 1 - \frac{A}{r} \tag{2}$$

is used. The work is aimed at analyzing the development of the process at the time moment, when the energy of a point explosion is equal to the energy of a burned gas, $r = R_x$, but provided that $R_x \neq \infty$. In other words, we intended to study the initial stage of detonation in reacting gas media by determining the scalar value of detonation wave velocity. The transition to the Chapman-Jouguet regime begins at some distance R_x from the center [4, 5], when the energy of the system appreciably increases. The model supposes that the pressures at the front and in the explosion region are equal, which results in the appearance of a high temperature at the transition point behind the shock front, since the main part of the substance mass is concentrated in a thin layer of the blast wave.

A lot of researches were devoted to the emergence of detonation in a hydrogen-oxygen mixture [6–8]. At the same time, there are no works in the scientific literature dealing with the influence of initial conditions on the detonation process, when the blast wave propagates in the gas environment. In this paper it was possible to obtain the necessary results by studying the chain reactions [9] of the interaction of hydrogen with oxygen.

2. Determination of the detonation wave velocity in an explosive gas mixture

2.1 Explosion in a chemically inert gas mixture

Consider an explosion in a chemically inert gas mixture. Let the point explosion occur instantly in a perfect gas with density ρ_0, and a shock wave propagates in the gas from the point of energy release. We intend to analyze the initial stage of the process of shock wave propagation, when the shock wave amplitude is still so high that the initial gas pressure, P_0, can be neglected. This assumption is equivalent to a neglect of the initial internal gas energy in comparison with the explosion one, i.e., we consider a strong explosion. The problem is to determine the velocity of the blast wave, when the wave front is modeled by a rigid piston compressing the volume of the gas in front (**Figure 1**). The main regularities of the process are well-known [10], and there is a simple approximate method to find them.

Let the total mass of a gas engaged into a blast wave be concentrated in a thin layer near the front surface. The gas density here is constant and equal to that at the front,

$$\rho_1 = \frac{\gamma + 1}{\gamma - 1}\rho_0. \tag{3}$$

This formula can be derived from the formula for strong shock waves [11] in the case where the Mach number $M \gg 1$. To avoid a misunderstanding, note that, in this case, we mean a transformation of a medium denoted by subscript. 0 (the medium at rest before the explosion) into a medium denoted by subscript 1. The layer thickness Δr is determined from the condition of mass conservation,

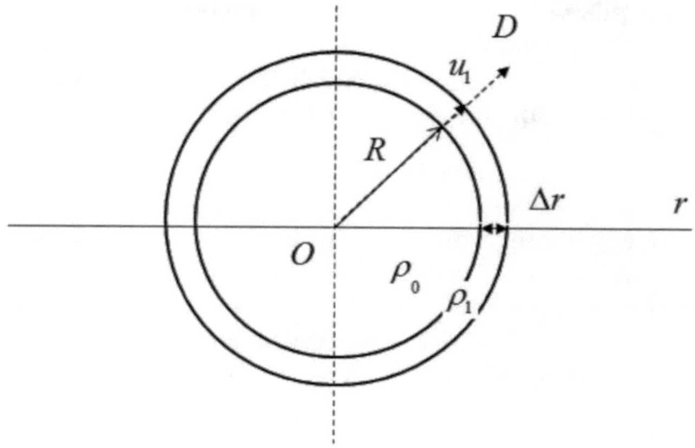

Figure 1.
Schematic diagram of a shock wave from the point explosion.

$$4\pi R^2 \Delta r \rho_1 = \frac{4}{3}\pi R^3 \rho_0; \tag{4}$$

whence

$$\Delta r = \frac{R\rho_0}{3\rho_1} = \frac{R}{3}\left(\frac{\gamma - 1}{\gamma + 1}\right). \tag{5}$$

Since the layer is very thin, the gas velocity in it almost does not change and coincides with that at the front,

$$u_1 = \frac{2D}{\gamma + 1}. \tag{6}$$

In the shock wave theory, a more accurate formula is considered, which couples the gas flow velocity behind the shock wave front, u_1, with the front velocity D: $u_1 = \frac{2D}{(\gamma+1)\left[1-\left(\frac{b_0}{D}\right)^2\right]}$, where b_0 is the sound velocity in the unperturbed gas. The gas mass in the layer is finite and equal to the mass m of the gas originally contained within a sphere of radius R,

$$m = \frac{4}{3}\pi R^3 \rho_0. \tag{7}$$

Let us denote the pressure at the inner layer side as P_c, and let it equal α times the pressure at the wave front, $P_c = \alpha P_1$. Newton's second law for the layer Δr in thickness reads

$$\frac{d}{dt}(mu_1) = 4\pi R^2 P_c = 4\pi R^2 \alpha P_1. \tag{8}$$

It can be used only within the limits

$$0 < R \le R_x^0, \tag{9}$$

where the quantity R_x^0 is determined from energy considerations. At this point, the kinetic energy of the blast wave is still high enough, and its velocity considerably exceeds that of sound in the unperturbed gas medium.

Here, we arrive at a detailed mathematical representation of formulas and equations, by using the already known relations (4) and (8), which are the conservation laws for the mass and the moment, respectively. However, the latter are not enough for the problem to be solved. One more equation is needed,

$$E = E_T + E_k = const; \tag{10}$$

which is the energy conservation law. The explosion energy is constant and equal to a sum of two terms: the potential, E_T, and kinetic, E_k, energies. The general system consists of three equations—Eqs. (4), (8), and (10)—appended by the condition of strong explosion, $M \gg 1$, when formula (6), the relations (4), (8), (10) and the condition of strong explosion

$$M \gg 1 \tag{11}$$

are valid. Moreover, we have Eq. (6), and

$$P_1 = \frac{2}{\gamma + 1} \rho_0 D^2, \tag{12}$$

where P_1 is the pressure at the front of the shock wave. Formula (12) follows from the relation $\frac{P_1}{P_0} = \frac{2\gamma M^2 - \gamma + 1}{\gamma + 1}$ of work [11] in the case where $M \gg 1$: $\frac{P_1}{P_0} \approx \frac{2\gamma M^2}{\gamma + 1} = \frac{2\gamma \times \frac{D^2}{b_0^2}}{(\gamma+1)} = \frac{2\gamma \times \frac{D^2 \rho_0}{\gamma P_0}}{(\gamma+1)} = \frac{2\rho_0 D^2}{P_0(\gamma+1)}$.

It should be noted that, in the given system of equations, relation (4) does not determine a connection between the regions separated by the shock wave front (regions 0 and 1). Instead, it couples the states before the explosion and after it. While solving this problem for the one-dimensional centrally symmetric flow, we come back to Eq. (8).

The mass itself depends on the time, so that it is the momentum mu_1 rather than the velocity that should be differentiated with respect to the time. The mass is subjected to the action of the force $4\pi R^2 P_c$ directed from the inside, because the pressure P_c is applied to the inner side of the layer. The force acting from the outside is equal to zero, because the initial pressure of the gas is neglected. By expressing the quantities u_1 and P_1 in Eq. (8) in terms of the front velocity $D = \frac{dR}{dt}$ and using formulas (6) and (12), we obtain the new relation

$$\frac{1}{3} \frac{d}{dt} R^3 D = \alpha D^2 R^2. \tag{13}$$

Bearing in mind that

$$\frac{d}{dt} = \frac{d}{dR} \frac{dR}{dt} = D \frac{d}{dR} \tag{14}$$

and integrating Eq. (13), we find

$$D = aR^{-3(1-\alpha)} \tag{15}$$

where a is the integration constant. To determine the parameters a and α, let us take the energy conservation law into account. The kinetic energy of the gas is equal to

$$E_k = \frac{mu_1^2}{2}. \tag{16}$$

The internal energy is concentrated in a "cavity" confined by an infinitesimally thin layer. The pressure in the cavity is equal to P_c. Actually, this means that, strictly speaking, the whole mass is not contained in the layer. A small amount of the substance is included into the cavity as well. In gas dynamics, the specific internal energy of the ideal gas is calculated by the formula $e = \frac{P}{\rho}\left(\frac{1}{\gamma-1}\right)$, where P is the pressure, ρ the density, and γ the adiabatic index. Therefore, the internal energy is equal to

$$E_T = \frac{1}{\gamma - 1} \times \frac{4\pi R^3}{3} P_c, \quad (17)$$

so that

$$E = E_T + E_k = \frac{1}{\gamma - 1} \times \frac{4\pi R^3}{3} P_c + \frac{m u_1^2}{2}. \quad (18)$$

Expressing the quantities P_c and u_1 once more in terms of D and substituting $D = aR^{-3(1-\alpha)}$, we obtain

$$E = \frac{4}{3}\pi\rho_0 a^2 \left[\frac{2\alpha}{\gamma^2 - 1} + \frac{2}{(\gamma+1)^2}\right] R^{3-6(1-\alpha)}. \quad (19)$$

Since the explosion energy E is constant, the power exponent of the variable R must be equal to zero. This means

$$\alpha = \frac{1}{2}. \quad (20)$$

We determine the constant a from Eq. (19)

$$a = \left[\frac{3}{4\pi} \times \frac{(\gamma-1)(\gamma+1)^2}{3\gamma - 1}\right]^{\frac{1}{2}} \left(\frac{E}{\rho_0}\right)^{\frac{1}{2}}, \quad (21)$$

and substituting it together with Eq. (20) into formula (15), we arrive at the expression for the shock wave velocity in the case of point-like explosion

$$D = \left[\frac{3}{4\pi} \times \frac{(\gamma-1)(\gamma+1)^2}{3\gamma - 1}\right]^{\frac{1}{2}} \left(\frac{E}{\rho_0}\right)^{\frac{1}{2}} R^{-3/2}, \quad (22)$$

or

$$D = \xi_0 \left(\frac{E}{\rho_0}\right)^{\frac{1}{2}} R^{-3/2}, \quad (23)$$

where

$$\xi_0 = \left[\frac{3}{4\pi} \times \frac{(\gamma-1)(\gamma+1)^2}{3\gamma - 1}\right]^{\frac{1}{2}} = const. \quad (24)$$

2.2 Theory of explosion in a combustible mixture of gases

Distinctive features of the problem consist in that the exothermic chemical reactions are possible in such a medium. Therefore, it is quite reasonable to assume that the blast wave continuously transforms into the detonation one. Let us consider the following model. An explosion in the gas generates a strong shock wave, which propagates over the gas and heats it up to a state, in which burning reactions become probable. We denote the energy of explosion by E_0. The energy U released at the combustion of the gas is equal t

$$U = \frac{4}{3}\pi R_1{}^3 \rho_0 Q', \quad R_0 \ll R_1 \tag{25}$$

where Q' is the specific heat released in the medium (per mass unit of the medium). The process is considered at the time moment t_1, when $R = R_1$ (**Figure 2**). Supposing that $E_0 > U$, we determine a condition, under which the detonation energy weakly affects the gas flow [12],

$$R_1 < R_x, \tag{26}$$

where $R_x^3 = \frac{3E_0}{4\pi Q' \rho_0}$. Let the charge have a finite radius R_0 Then, when applying the conventional theory of point explosion to the description of the motion, we have to use the estimation

$$R_0 < R < R_x. \tag{27}$$

It should be noticed that conditions (26) and (27) strongly restrict the scope, where the laws of point explosion in an inert gas are applicable to the flows in the detonating medium. However, if the energy $\boldsymbol{E_0}$ is high, and if it is released in a small volume, the flow in the region $\boldsymbol{R_1 < R_x}$ would mainly occur as it does at an ordinary point explosion. On the other hand, for the time moment $\boldsymbol{t_2}$, at which

$$R = R_2 \text{ and } E_0 < U, \tag{28}$$

the combustion processes start to play a dominating role, and the gas flow will possess the main characteristics of the detonation combustion [12].

From the aforesaid, some interesting conclusions can be drawn.

1. The theory of a point explosion is proposed to be used for a combustible mixture of gases within the limits $R_0 < R < R_x$, if the proposed model of transformation of a blast wave into a detonation one is valid for the given

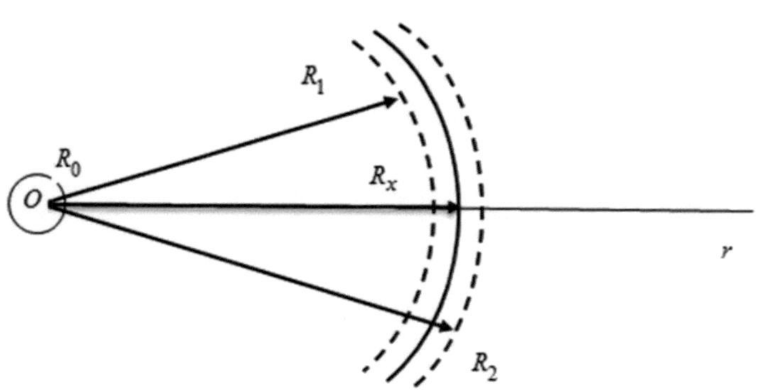

Figure 2.
Scenario of the continuous transformation of a blast wave into a detonation one: R_0 is the charge radius, R_x is the initial threshold, R_2 is the final threshold, where the transformation of the strong detonation mode into the Chapman-Jouguet one is possible.

mixture. Another scenario is probable, when the detonation is impossible under the given physico-chemical conditions in the gas medium, and the blast wave simply fades.

2. When $R \to R_x$, the energy of the system considerably changes, $E \neq const$, increasing almost twice as much, which has to be taken into consideration while studying the gas motion at this stage.

3. It is evident that, if $R \to R_x$, the energy becomes proportional to the cube of the sphere radius, $E \sim R^3$.

Hence, we come to an idea of that, for the model of point explosion in a combustible mixture of gases to be valid at $R \to R_x$ under our conditions, it should be either modified or extended. Look once more at formula (19) expressing the energy conservation law. In the theory of point explosion for a usual non-detonable mixture of gases, it is adopted that $E = const$, which results in $\alpha = \frac{1}{2}$. However, in the case $\alpha = 1$, Eq. (19) yields $E \sim R^3$, which is necessary in our case. One can see that the energy conservation law allows the following set of relations:

$$\alpha = 1; \tag{29}$$

$$E = \frac{4}{3}\pi\rho_0 a^2 \left[\frac{2}{\gamma^2 - 1} + \frac{2}{(\gamma + 1)^2}\right] R^3; \tag{30}$$

$$a = \left[\frac{3(\gamma - 1)(\gamma + 1)^2}{16\pi\gamma}\right]^{\frac{1}{2}} \left(\frac{E}{\rho_0}\right)^{\frac{1}{2}} R^{-3/2}. \tag{31}$$

Substituting the new values of $\boldsymbol{a}$ and $\boldsymbol{\alpha}$ into formula (15), we obtain

$$D = a = \left[\frac{3(\gamma - 1)(\gamma + 1)^2}{16\pi\gamma}\right]^{\frac{1}{2}} \left(\frac{E}{\rho_0}\right)^{\frac{1}{2}} R^{-3/2}. \tag{32}$$

According to the integration rules, the quantity a is a constant. Hence, a new formula for the velocity of a blast wave in the reacting gas medium is proposed:

$$D = \left[\frac{3(\gamma - 1)(\gamma + 1)^2}{16\pi\gamma}\right]^{\frac{1}{2}} \left(\frac{E}{\rho_0}\right)^{\frac{1}{2}} R^{-3/2} = const. \tag{33}$$

The law of conservation of energy gives unpredictable results, but these results are quite possible, given that the energy of the system is changing.

2.3 Formula for the velocity of a spherical wave

Let us determine the shock wave velocity in the critical zone, when $R \to R_x$ and $R \to R_2$ (**Figure 2**). One can consider a simplified version, when the transition occurs at a distance R_x from the center [4], but for the formation of normal detonation it is necessary to isolate the transition interval. As an example, let us consider the detonating gas, $2H_2 + O_2 = 2H_2O + Q$, where $Q = 286.5$ kJ/mol is the thermal effect obtained at a combustion of one mole of hydrogen. Let this reaction (an initial explosion) be initiated. The energy of the system is

$$E = Vn_{H_2}q + E_0, \tag{34}$$

where V is the volume of a certain ball, n_{H_2} the concentration of hydrogen molecules in it, q the thermal effect produced by one hydrogen molecule, and E_0 the initial energy of a charge of radius R_0 (recall that $R_0 \ll R_x$, but the current radius of the sphere $R \to R_x$). The volume of the ball and the concentration of hydrogen in it are calculated using the known formulas: $V = \frac{4}{3}\pi R^3$, $n_{H_2} = \frac{P_0}{K^* T_0} N_A c$, *where* $\frac{P_0}{K^* T_0} = \frac{\rho_0}{\mu}$; P_0, T_0, and ρ_0 are the initial pressure, temperature, and density of the gas mixture; K^* is the universal gas constant; N_A is the Avogadro constant; c the current content of hydrogen in the mixture (it is supposed that all the hydrogen burns out in the course of the reaction); and μ is the molar mass of the mixture. Hence,

$$E = \frac{4}{3}\pi R^3 \frac{P_0}{K^* T_0} N_A c q + E_0. \tag{35}$$

Substituting Eq. (35) into Eq. (33), we obtain

$$D = \left[\frac{(\gamma+1)^2(\gamma-1)N_A q c}{4\gamma\mu} + \left(\xi_0'\right)^2 \frac{E_0}{\rho_0 R^3}\right]^{\frac{1}{2}},$$

where

$$\xi_0' = \left[\frac{3(\gamma-1)(\gamma+1)^2}{16\pi\gamma}\right]^{\frac{1}{2}}. \tag{36}$$

At the time moment, when $\boldsymbol{R \to R_2}$, where $\boldsymbol{R_2 > R_x \gg R_0}$, the second term in the brackets tends to zero, $(\boldsymbol{\xi'_0})^2 \frac{\boldsymbol{E_0}}{\boldsymbol{\rho_0 R^3}} \boldsymbol{\to 0}$, whence we obtain

$$D = \left[\frac{(\gamma+1)^2(\gamma-1)Qc}{4\gamma\mu}\right]^{\frac{1}{2}}, \tag{37}$$

taking into account that $Q = N_A q$, where Q is the thermal energy of one hydrogen mole. The final formula (37) is suggested to describe the velocity of a detonation wave. Above the threshold R_2, the charge energy E_0 loses its importance; further, the energy of the system is replenished only by the first term Eq. (35), which demonstrates the real wave velocity. Provided that formula (37) is valid, the examined quantity does not depend on the mixture pressure. At the initial time moment, the velocity is constant, and it is governed by the following parameters: the combustion energy per one mole of the combustible gas, Q; the fraction of the burned-out gas, c; the molar mass of the mixture, μ; and the adiabatic index for the given mixture of gases, γ.

For a plane wave, the following formula is widely known [11, 13]:

$$D = \sqrt{2(\gamma^2 - 1)Q^*} \tag{38}$$

where Q^* is the ratio between the energy released by a substance to the mass flow of this substance. As a result, by comparing formulas (37) and (38), we come to a conclusion that they are very similar, although the former seems to be more

Gas mixture	D_s [m/s]	D_n [m/s]	ε [%]
66.6%H_2+33.3%O_2	2550	2830	9.9
25%C_2H_2+75%O_2	2089	2330	10.3

Table 1.
Shock wave velocities.

adequate for the description of the spherical detonation at the beginning of the process. The results of calculations for two different gas mixtures are compared in the **Table 1**, where D_s is the velocity of a spherical wave calculated by the new formula (37) at the beginning of the detonation, when $R = R_2$; D_n is the plane wave velocity at the final stage of detonation, when $R \to \infty$, taken from work [14]; and ε is the corresponding relative difference.

In this work, the ideal case of the transformation of an explosive spherical wave into the Chapman-Jouguet mode is considered. From this viewpoint, formulas (33) and (37) prove that the regime of normal spherical detonation can exist at the beginning of the process, much earlier before the curvature radius can be assumed tending to the infinity. Moreover, it demonstrates a possibility of the existence of the normal spherical detonation with a lower velocity of a shock wave in comparison with the classical one. The mathematical expression (38) is "actual" at the final stage, when the radius tends to infinity, i.e. for the plane wave. It should be noted that, in the gas dynamics researches, instead of the shock wave velocity, its ratio to the sound velocity in the unperturbed gas medium, b_0, i.e. the Mach number M, is often used,

$$M = \frac{D}{b_0}. \tag{39}$$

With regard for the formula for the sound velocity,

$$b_0 = \sqrt{\gamma \frac{P_0}{\rho_0}} = \sqrt{\gamma \frac{K^* T_0}{\mu}}, \tag{40}$$

and expression (37), we obtain

$$M = \left[\frac{(\gamma + 1)^2 (\gamma - 1) Qc}{4\gamma^2 K^* T_0} \right]^{\frac{1}{2}}. \tag{41}$$

Formula (41) demonstrates the dependence of the Mach number on the adiabatic index γ, the combustion heat Q, the fraction of the burned-out gas c, and the temperature of the medium T_0. By varying those quantities, it is possible to regulate the shock transition intensity.

3. Conditions for the appearance of spherical detonation in the interaction of hydrogen with oxygen

3.1 Some issues concerning the chemical reaction kinetics

The process of shock wave propagation is very fast. For instance, at the shock wave velocity $D = 2500$ m/s and the gas layer thickness $r = 0.005$ m the shock compression of the substance lasts $t = 2 \times 10^{-6}$ s. This means that the dominant part of a compressed substance must react within such a short time interval; only in this case, we may talk about the supersonic burning as a self-supporting process [13]. Proceeding from this viewpoint, let us consider some issues of the kinetics of the chemical reaction of H_2 and O_2.

First of all, it should be emphasized that the matter concerns chain reactions. The Haber scheme [9] and the development of a chain reaction with the Haber cycle look like

$$OH + H_2 = H_2O + H, \tag{42}$$

$$H + O_2 + H_2 = H_2O + OH, \tag{43}$$

$$H + O_2 = OH + O, \tag{44}$$

$$O + H_2 = OH + H, \tag{45}$$

$$OH + OH = H_2O_2 \rightarrow \text{chain break}, \tag{46}$$

$$H + H = H_2 \rightarrow \text{chain break}, \tag{47}$$

$$H + \text{wall} \rightarrow \text{chain break}, \tag{48}$$

$$OH + \text{wall} \rightarrow \text{chain break}. \tag{49}$$

Reactions (42) and (43) correspond to the chain continuation, reactions (44) and (45) to the chain branching, and reactions (46)–(49) to the chain break. For reaction (42), the corresponding activation energy is supposed to be high, with not every collision of OH and H_2 resulting in the reaction between them. On the contrary, reaction (43) runs at every ternary collision [9]. The cycle of reactions (42) and (43) composes a repeating chain link. According to Haber, 5–10, on the average, cycles must pass before reaction (44) occurs and there emerges a branching in the chain. Let us consider reactions (43) and (44), which compete with each other. Denoting the rate of reaction reaction (44) as W_3 and that of reaction (43) as W_2, the probability of branching δ can be defined as the rate ratio

$$\delta = \frac{W_3}{W_2}. \tag{50}$$

In Semenov's book [9], the expression for δ is given as

$$\delta = \frac{2.5 \times 10^5 \exp\left(\frac{-E_3}{K^* T_2}\right)}{[H_2]}, \tag{51}$$

where $[H_2]$ is the partial pressure of hydrogen in units of mm Hg (the numerical coefficient of 2.5×10^5 in the nominator is multiplied by 1 mm Hg; therefore, the pressure in the denominator is expressed in terms of mm Hg units), E_3 is the activation energy of reaction (44), K^* the gas constant, and T_2 the medium temperature (in Kelvin degrees). According to Semenov's data [9, 15], $E_3 = 16$ kcal/mol. Formula (51) shows that δ strongly depends on the temperature, so that the process can be substantially accelerated as the temperature grows. Moreover, it turns out that the cycle of reactions (42) and (43) with branching (44) does not describe the fastest mechanism. There may exist a case where

$$W_3 = W_2, \tag{52}$$

or

$$\delta = 1. \tag{53}$$

From the physical viewpoint, this means that the probability reaches the maximum, and the branching occurs at every chain link. Then the interaction scheme changes, reaction (44) substitutes reaction (43), and a transformation to the Lewis scheme takes place. In this case, we obtain $OH + H_2 = H_2O + H$, $H + O_2 = OH + O$, and so on, i.e. the temperature T_x, at which $\delta = 1$, is a critical one, when the kinetics of the interaction between hydrogen and oxygen undergoes qualitative changes. Let us write down the Lewis scheme in the complete form [9],

$$OH + H_2 = H_2O + H, \tag{54}$$

$$H + O_2 = OH + O, \tag{55}$$

$$OH + H_2 = H_2O + H, \tag{56}$$

$$O + H_2 = OH + H, \tag{57}$$

$$H + \text{wall} \rightarrow \text{chain break}, \tag{58}$$

$$O + \text{wall} \rightarrow \text{chain break}, \tag{59}$$

$$OH + \text{wall} \rightarrow \text{chain break}. \tag{60}$$

In the summarized form, the cycle reaction looks like

$$OH + 3H_2 + O_2 = OH + 2H + 2H_2O, \tag{61}$$

and just this reaction is associated with the first fastest initial chain transformations that give rise to detonation.

3.2 Medium state at the shock wave front. Critical temperature

Let a point explosion took place in a gas medium. In our case, the matter concerns the reacting gas media; therefore, the blast wave extinction may occur more slowly that usually; or it can be absent altogether, because a strong mechanism of chain reactions between hydrogen and oxygen starts to play its role. The ultimate result depends on the physico-chemical properties of the gas mixture and the initial energy of explosion. From this point of view, the most interesting is the model of a transition of the strong (overcompressed) detonation into the Chapman-Jouguet regime.

The shock wave propagates from a region with a higher pressure into a region where the pressure is lower. The gas dynamics usually considers waves that have a sharp front. The region of shock-induced transition is a discontinuous surface, the shock wave front. The unperturbed state is designated by subscript 1 and the perturbed one by subscript. 2. The density ρ, pressure P, and temperature T change in a jump-like manner across the front. The relations between the parameters (P_1, T_1, ρ_1) and (P_2, T_2, ρ_2) follow from the Hugoniot relations (the conservation laws) and the equation of ideal gas [11]. It is known that

$$\frac{\rho_2}{\rho_1} = \frac{(\gamma + 1)M^2}{2 + (\gamma - 1)M^2}; \tag{62}$$

$$\frac{P_2}{P_1} = \frac{2\gamma M^2 - \gamma + 1}{\gamma + 1}; \tag{63}$$

$$\frac{T_2}{T_1} = \frac{(2\gamma M^2 - \gamma + 1)(2 + (\gamma - 1)M^2)}{(\gamma + 1)^2 M^2}, \tag{64}$$

where Mach number, M , $\gamma = \frac{C_p}{C_V}$ is the adiabatic exponent (for a two-atom ideal gas, γ =1.4 [16]). In such a manner, when a shock wave propagates in gases, we should consider the medium near (subscript 1) and at the front (subscript 2). To characterize the latter, we must know an important parameter, the shock wave velocity or the Mach number. In our case, using expression (37) and the formula for sound velocity (40), we obtain (41).

Now, let us carry out a simple gedanken experiment. Let a spherical reactor contain a hydrogen-oxygen mixture with the initial parameters (P_0, $T_0 = 293$ K). Let us heat up the mixture to the temperature $T_1 < T_1^*$, where T_1^* is the ignition temperature of the static medium. We initiate a reaction using an explosion and should observe a continuous transformation of a blast wave into the detonation of the hydrogen-oxygen mixture. At the wave front, the medium parameters are (P_2, T_2). Let

$$T_2 = T_x, \tag{65}$$

i.e. the critical temperature T_x is attained, and the reaction develops, being driven by the chain reaction mechanism according to the Lewis scheme. In order to determine the critical temperature T_x, let us use formula (51). Taking into account that $\delta = 1$, we obtain the transcendental equation for the critical temperature T_x

$$\frac{2.5 \times 10^5 \exp\left(-\frac{E_3}{K^* T_x}\right)}{[H_2]} = 1, \tag{66}$$

where

$$[H_2] = cP_2 \tag{67}$$

is the partial hydrogen pressure (in mm Hg units) at the shock wave front [17], P_2 is the total pressure in the mixture (in mm Hg units) at the shock wave front, and c the hydrogen content in the mixture (coefficient). With regard for Eq. (67), we obtain

$$\frac{2.5 \times 10^5 \exp\left(-\frac{E_3}{K^* T_x}\right)}{[cP_2]} = 1. \tag{68}$$

Let us express P_2 in the denominator of Eq. (68) in terms of known quantities. Before the reaction started (the initiation of the explosion), the gas mixture pressure was P_0, and its temperature was $T_0 = 293$ K. As the mixture is heated up to T_1, its pressure increases to

$$P_1 = P_0 \frac{T_1}{T_0}. \tag{69}$$

From Eq. (63), it follows that

$$P_2 = \frac{2\gamma M^2 - \gamma + 1}{\gamma + 1} P_1, \tag{70}$$

or, in view of Eq. (69),

$$P_2 = \frac{(2\gamma M^2 - \gamma + 1) P_0 T_1}{(\gamma + 1) T_0}. \tag{71}$$

The temperature T_1 in formula (71) is expressed in terms of T_x and the Mach number M as follows:

$$\frac{T_x}{T_1} = \frac{\left(2\gamma M^2 - \gamma + 1\right)\left(2 + (\gamma - 1)M^2\right)}{(\gamma + 1)^2 M^2}. \tag{72}$$

Whence, we obtain

$$T_1 = \frac{(\gamma + 1)^2 M^2 T_x}{\left(2\gamma M^2 - \gamma + 1\right)\left(2 + (\gamma - 1)M^2\right)}, \tag{73}$$

or, taking Eq. (73) into account,

$$P_2 = \frac{(\gamma + 1)M^2 T_x P_0}{T_0\left(2 + (\gamma - 1)M^2\right)}. \tag{74}$$

The denominator in formula (68) also includes the hydrogen content, c. If we assume that all hydrogen in the gas mixture burns out, we can express c using the Mach number (Eq. (41)) and the temperature of the gas medium T_1,

$$c = \frac{4\gamma^2 M^2 K^* T_1}{(\gamma - 1)(\gamma + 1)^2 Q}, \tag{75}$$

or, in accordance with Eq. (73),

$$c = \frac{4\gamma^2 M^4 K^* T_x}{(\gamma - 1)\left(2\gamma M^2 - \gamma + 1\right)\left(2 + (\gamma - 1)M^2\right)Q} \tag{76}$$

(in this case, we impose a restriction on the gas mixture composition, $0 < c \leq 0.66$). From Eq. (67) and using Eqs. (76) and (74), we obtain the partial pressure of hydrogen at the shock wave front,

$$[H_2] = \frac{4\gamma^2(\gamma + 1)M^6 K^* P_0}{(\gamma - 1)\left(2\gamma M^2 - \gamma + 1\right)\left(2 + (\gamma - 1)M^2\right)^2 Q T_0} T_x^2. \tag{77}$$

Then, formula (66) reads

$$T_x^2 = \frac{2.5 \times 10^5 Q T_0 (\gamma - 1)\left(2\gamma M^2 - \gamma + 1\right)\left(2 + (\gamma - 1)M^2\right)^2}{4\gamma^2(\gamma + 1)M^6 K^* P_0} \exp\left(-\frac{E_3}{K^* T_x}\right). \tag{78}$$

Hence, we obtained the dependence which connects the initial pressure in the medium and the Mach number with the critical temperature at the shock wave front.

3.3 Results and discussion

After the substitution of the corresponding numerical values of physical parameters of the hydrogen-oxygen mixture and taking into account that $\gamma = 1.4$, $Q = 286.5$ kJ/mol, $K^* = 8.31$ J/(mol K), $E_3 = 16 \times 10^3 \times 4.19$ J/mol, and $T_0 = 293$ K, Eq. (78) reads

$$T_x{}^2 = \frac{5.38 \times 10^{10} (2 + 0.4M^2)^2 (2.8M^2 - 0.4)}{P_0 M^6} \exp\left(-\frac{8067}{T_x}\right). \tag{79}$$

Most of these quantities are well known. The values of the others are chosen for practical reasons. So, for example, $T_0 = 293$ K, this is the temperature at which the experimental setup operates. The most optimal for laboratory conditions is the pressure of the gas mixture $P_0 = 60$ mm Hg, since at $P_0 > 60$ mm Hg the shock wave acquires destructive energy. Using expression (79), let us calculate the critical temperature for two Mach numbers, (i) $M = 2.15$ and (ii) $M = 4.78$, i.e. for shock waves of two types, but at the fixed initial pressure $P_0 = 60$ mm Hg. Experimental data indicate that detonation is not observed at $M < 2.15$, weak shock waves become waves of compression and rarefaction. At the same time, the value $M = 4.78$ was selected as the largest one obtained from expression (41) at the following parameters: $c = 0.66$, $T_1 = T_0 = 293$ K, and $\gamma = 1.4$. In the first case ($M = 2.15$ and $P_0 = 60$ mm Hg)

$$T_x{}^2 = 1.69 \times 10^9 \exp\left(-\frac{8067}{T_x}\right). \tag{80}$$

In the second one ($M = 4.78$ and $P_0 = 60$ mm Hg),

$$T_x{}^2 = 5.93 \times 10^8 \exp\left(-\frac{8067}{T_x}\right). \tag{81}$$

The transcendental equations were solved with the use of the software package "Consortium Scilab (Inria, Enpc)" with the program code "Scilab-4.1.2". After the corresponding calculations, we obtained $T_x = 1120$ K for the first case and $T_x = 1420$ K for the second one. The interval of researches can be expanded to determine the critical temperatures for Mach numbers within line segment [2; 5] with an increment of 0.2. Only real-valued roots, which have a physical sense, must be taken into consideration. The corresponding plot for the dependence $T_x \sim f(M)$ at $P_0 = 60$ mm Hg is shown in **Figure 3**.

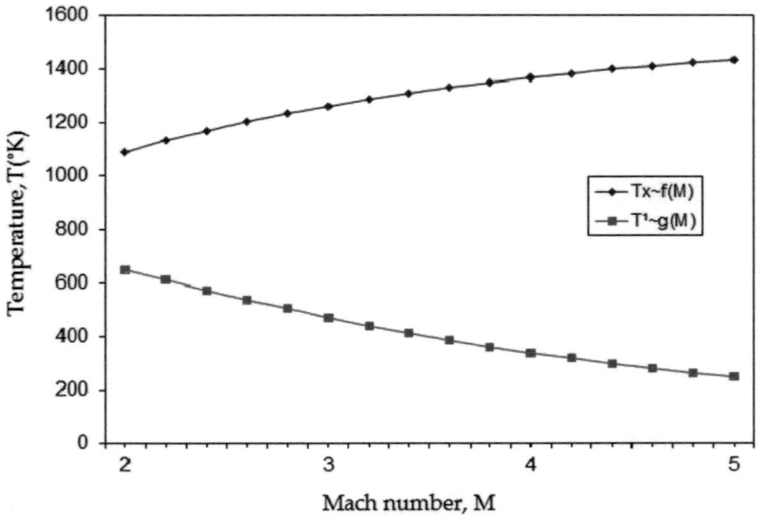

Figure 3.
Dependences of the critical temperature at the shock wave front, $T_x \sim f(M)$, and the temperature of static medium $T^1 \sim g(M)$, at which the detonation is possible, on the Mach number at a fixed pressure $P_0 = 60$ mm Hg.

One can see that the critical temperature grows nonlinearly with the Mach number. This behavior is not of surprise. The stronger the shock wave, the higher is the pressure at its front, and the higher is the probability of the chain break. As a result, we obtain the critical temperature growth, because the probability of chain break can be compensated only by the probability of chain branching, which increases with the medium temperature. However, this is not the main point. Knowing the critical temperature and the Mach number, it is possible to determine the initial temperature of the gas medium required for the detonation to take place. In other words, it is possible to determine such a temperature T^1 in front of the shock wave front that the corresponding wave would stimulate the detonation. Using expression (73), let us plot the dependence of T^1 on M at $P_0 = 60$ mm Hg (**Figure 3**). It allows us to determine the initial temperature of the medium, at which the detonation of the gas mixture becomes possible for the given Mach number. Moreover, in accordance with Eq. (41), the initial temperature and the known Mach number determine the hydrogen content. Hence, the critical temperature is unambiguously related with the Mach number and, therefore, with the initial parameters of the hydrogen-oxygen mixture.

4. Conclusions

To summarize, it should be noted that formula (37) determines the velocity of a detonation wave at the initial stage, if this wave is generated at the combustion of some “portion” (the parameter c) of a combustible gas, when $R \to R_2$ (see **Figure 2** and the model described the transformation of a blast wave in a detonation one). This formula is valid for spherical wave, in contrast to formula (38) known from the literature, which was obtained for plane waves. Thus, provided that the shock wave velocity or the Mach number is known, the solution of one of the basic gas dynamics problems can be obtained, i.e. we can find the parameters (P_1, T_1, ρ_1) at the wave front, if we know the set (P_0, T_0, ρ_0) of parameters for the unperturbed medium. In particular, the determined parameters are necessary for studying the kinetics of a chemical reaction in the course of the shock transition. There is no doubt that the stoichiometric mixture of hydrogen and oxygen will generate a detonation wave. However, it is difficult to assert the same for the mixture with 12% of hydrogen. In this case, it is necessary to consider the reaction mechanism itself.

The expression (66) obtained for the critical temperature is the simplest criterion for the transformation of the blast wave into a detonation one. At this temperature, $T_2 = T_x$ if $\delta = 1$, i.e. the probability of branching becomes maximum for the scheme of chain reactions with the hydrogen-oxygen interaction, which was considered above. The obtained Eq. (78) allows the sought value to be determined as a function of the Mach number, provided that the initial pressure is fixed. The critical temperature is the threshold of a detonation in a gas mixture, because the supersonic burning is impossible at temperatures below it. For example, let us analyze gas mixtures with different hydrogen contents of 66.6, 60, and 50%, and the temperature of static medium $T_1 = 273$ K (**Table 2**).

Gas mixture	T_1 [K]	M	T_2 [K]	T_x[K]
$66.6\%H_2 + 33.3\%O_2$	273	4.95	1558	1427
$60\%H_2 + 40\%O_2$	273	4.72	1438	1420
$50\%H_2 + 50\%O_2$	273	4.31	1241	1390

Table 2.
Parameter changes at the shock transition ($T_1 = 273$ K and $P_0 = 60$ mm Hg).

Gas mixture	T_1 [K]	M	T_2 [K]	T_x[K]
$66.6\%H_2 + 33.3\%O_2$	373	4.23	1648	1384
$60\%H_2 + 40\%O_2$	373	4.04	1523	1365
$50\%H_2 + 50\%O_2$	373	3.70	1339	1339

Table 3.
Parameter changes at the shock transition ($T_1 = 373$ K and $P_0 = 60$ mm Hg).

Comparing the T_2- and T_x-values, we come to a conclusion that the process, which is of interest for us, occurs only in the first two cases. However, it is enough to raise the initial temperature $T_1 = 373$ K for the detonation to become possible at lower hydrogen concentrations (**Table 3**).

The practical results testify that the conditions for the emergence of spherical detonation have a drastic dependence on the temperature and the mixture composition. The relation obtained in this work allows the critical values of those parameters to be determined and, in such a manner, to stimulate the regime of supersonic burning in the hydrogen-oxygen mixture. In the future, the presented results can be used to determine the range of admissible values of the parameters of the hydrogen-oxygen mixture necessary for detonation. Thus, it becomes possible to improve the performance of engines, make them more efficient.

Nomenclature

Basic designations

D	shock wave velocity, detonation velocity
R_x	critical radius
M	Mach number
P, T, ρ	pressure, temperature, density of the medium
u_1	gas velocity behind the shock front
b_0	speed of sound in a stationary gaseous medium
γ	adiabatic index
E_0	explosion energy
U	burnt gas energy
Q	combustion energy of one mole of combustible gas
μ	molar mass
K^*	universal gas constant
N_A	Avogadro number
c	coefficient of flammable gas content in the mixture
W	chemical reaction rate
δ	branching probability
$[H_2]$	is the partial pressure of hydrogen
T_x	critical temperature
H_2	hydrogen molecule
O_2	oxygen molecule
H_2O	water molecule
O	oxygen atom
H	hydrogen atom
OH	compound of an oxygen atom with a hydrogen atom
$f(x)$	function of the variable x
$\exp(x)$	exponential function

Δx	increment of variable x
$\frac{d}{dx}$	derivative
$\ll$	much less
$\gg$	much more
$\rightarrow$	aspires to
$[a; b]$	line segment

Author details

Myron Polatayko
Kyiv National University, Kyiv, Ukraine

*Address all correspondence to: pmm.miron@gmail.com

References

[1] Orlenko LP, editor. Physics of Explosion. 3rd ed. Vol. 1. Moscow: Fizmatlit; 2004. 832 p

[2] Orlenko LP. Physics of Explosion and Impact: Tutorial. Moscow: Fizmatlit; 2006. 304 p

[3] Ch M. Numerical Modeling of Detonation. 1st ed. Berkeley: University of California Press; 1979. 485 p

[4] Polatayko MM. Determination of detonation wave velocity in an explosive gas mixture. Ukrainskii Fizicheskii Zhurnal. 2012;**57**(6):606-611

[5] Polatayko MM. Research of Spherical Detonation in Gases: A Collection of Articles. Saarbrucken: LAP LAMBERTAcademic Publishing; 2014. 70 p

[6] Fedorov AV, Tropin DA, Bedarev IA. Mathematical modeling of the suppression of the detonation of a hydrogen-oxygen mixture by inert particles. Fizika Gorenia i Vzryva. 2010; **46**(3):103-115

[7] Liberman MA, Ivanov MF, Kiverin AD, et al. On the mechanism of the transition of slow combustion to detonation in a hydrogen-oxygen mixture. Zhurnal Eksperimental'noi i Teoreticheskoi Fiziki. 2010;**138**(10): 772-788

[8] Khitrin LN. The Physics of Combustion and Explosion. Jerusalem: Israel Program for Scientific Translations; 1962. 448 p

[9] Semenov NN. Chemical Kinetics and Chain Reactions. Oxford: Clarendon Press; 1935. 480 p

[10] Zel'dovich YB, Raizer YP. Physics of Shock Waves and High-Temperature Hydrodynamic Phenomena. New York: Dover; 2002. 944 p

[11] Sysoev NN, Shugaev FV. Shock Waves in Gases and Condensed Media. Moscow: Moscow State University; 1987. 133 p

[12] Korobeinikov VP. Problems in the Theory of Point Explosions in Gases. Providence, RI: American Mathematical Society; 1976. 317 p

[13] Fickett W. Introduction to Detonation Theory. University of California: Berkeley; 1985. 216 p

[14] Chernyi GG. Gas Dynamics. Moskow: Nauka Publ; 1988. 424 p

[15] Semenov NN. Some Problems of Chemical Kinetics and Reactivity. Vol. 1. Princeton: Princeton Univ. Press; 1959. 254 p

[16] Matveev AN. Molecular Physics. Moscow: Mir Publishers; 1985. 450 p

[17] Ravdel AA, Ponomareva AM, editors. A Brief Handbook of Physico-Chemical Values. 8th ed. Leningrad: Khimiya; 1983. 232 p

Chapter 6

DNS for Turbulent Premixed Combustion

Dipal Patel and Martin Agelin-Chaab

Abstract

Most of practical combustion occurs in turbulent flows which involve strong coupling between turbulence and chemical processes. The heat release from combustion alters the fluid properties such as density and viscosity and in turns affects the turbulence. Direct numerical simulations (DNS) provides a tool for obtaining both temporally and spatially resolved data in three dimension (3D). This chapter presents a brief overview of importance of DNS in turbulent combustion, the role of turbulence and identifies different combustion modes. The mathematical formulation and numerical implementation for DNS are introduced. The second half of this chapter presents DNS results for ignition in both homogeneous and stratified mixtures. It has been found that minimum ignition energy is required to obtain successful ignition in different turbulence regimes. An increase in turbulent velocity fluctuation may leads to a misfire. Additionally the difference between growing flames and those which are quenched by turbulence have been discussed with the help of the reaction–diffusion balance analysis. Furthermore, the turbulence intensity and length scale of the mixture inhomogeneity have important influences on achieving self-sustained combustion following successful ignition events.

Keywords: premixed combustion, turbulent premixed regimes, Kolmogorov scale, flame structure, ignition

1. Introduction

Many different numerical methods have been developed for the solution of fluid flow problems. Many commercial computational fluid dynamics (CFD) codes are available and have become standard engineering tools for simulation of non-reacting flows. However, for combustion, CFD techniques are not well developed for its accuracy and robustness. The moment we introduce combustion in CFD, it invites additional complexities for stable, accurate and efficient reacting flow numerical solvers. Many standard CFD techniques are available and serve as basic methods for solving combustion problems. Reader can refer to many standard textbooks on the subject [1, 2].

Direct numerical simulations (DNS) is a CFD tool which resolves all flow features explicitly and is widely adopted in combustion research. The feasibility and challenges of DNS in tackling the problems of turbulent combustion is discussed in great detail by Cant [3]. DNS often demands a very large computational power, especially when resolving the forced ignition process of turbulent reacting flows. Despite the computation cost, DNS is highly accurate and provides an enormous amount of detailed information in comparison to experiments because it is either extremely expensive or impossible to obtain three-dimensional temporally and spatially resolved data by experimental means.

2. Why DNS?

There are three main strategies adopted in CFD simulations of turbulent flows. These strategies are:

- Reynolds averaged Navier–Stokes (RANS)
- Large eddy simulation (LES)
- Direct numerical simulation (DNS)

The philosophies of the above mentioned strategies can be understood from **Figure 1**, which illustrates the grid spacing requirements for DNS, RANS, and LES in relation to the turbulent kinetic energy spectrum $E(\kappa)$. In the RANS, the grid spacing (let us say Δx) is of the order of the inverse of energy containing wave number (i.e. $\Delta x \sim 1/l$), which tells us that the whole of the turbulent kinetic energy is unresolved in the context of RANS. In the LES, the grid size is close to the filter width Δ, which tells us that the physical processes are taking place for the wave number $\kappa < 1/\Delta$ are fully resolved, however, the physics at $\kappa > 1/\Delta$ are happening at subgrid level and thus remains unresolved. In the case of DNS, almost all the turbulent kinetic energy is resolved as the grid size is of the order of inverse of dissipation wave number. **Figure 2** shows the volume rendered of burnt products (blue), flame surface (red) and fresh reactants (transparent) in the computational cubic domain with comparison across three strategies. RANS only shows the statistical information of turbulence scales, whereas DNS provides time dependent instantaneous full information on turbulence scales with huge Reynolds number dependency.

In DNS turbulent fluid motion is simulated without any kind of physical approximation which indicates that you do not need any turbulence model for DNS,

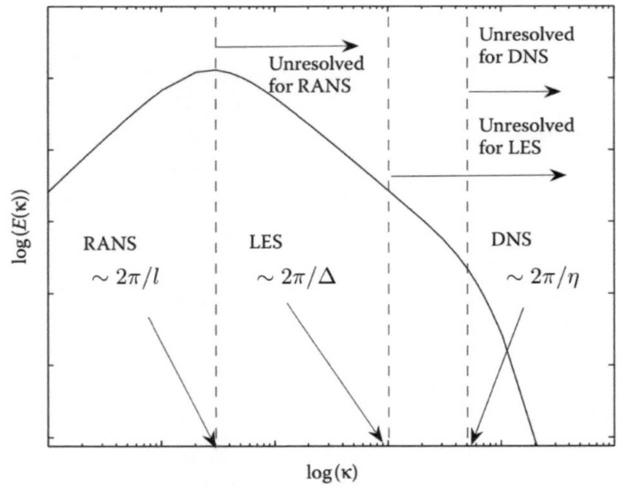

Figure 1.
The grid spacing for RANS, LES, and DNS on turbulent kinetic energy spectrum.

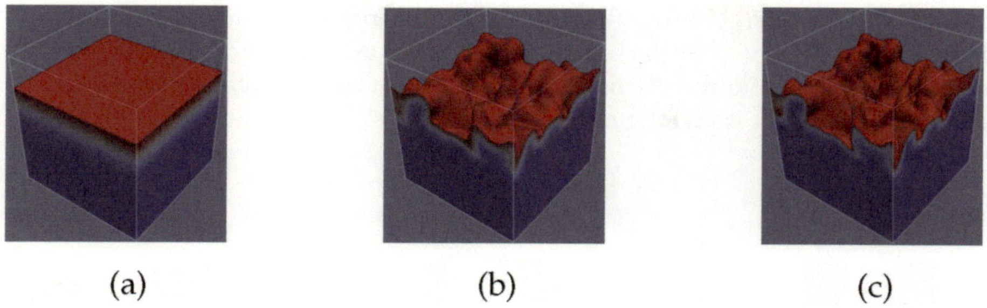

Figure 2.
Volume rendered view of reactants and products for (a) RANS, (b) LES, and (c) DNS.

all the length, time and velocity scales of turbulent flow are adequately resolved with the help of computational grid and time step used for given simulation case. It is important that the grid size (let us say Δx) in DNS needs to be smaller than the smallest significant length scale of turbulence, which is the Kolmogorov length scale η. Additionally, it is also important to have DNS computational domain that contains number of integral length scales so that enough number of large eddies are available to extract meaningful statistics. DNS simulations should be carried out for a number of integral time scales to have turbulent statistics that is independent of initial velocity field. The time step size for DNS should be smaller than the smallest time scale of turbulence. The computational time can be estimated by the product of grid points and number of time steps. This makes DNS extremely computationally expensive in nature. However, for case of compressible flow, computational time for DNS depends on Courant number, acoustic velocity, and Mach number.

3. Turbulent combustion

Combustion requires fuel and oxidiser to mix at the molecular level. How this takes place in turbulent combustion that depends on the turbulent mixing process. The general view is that once a range of different size eddies has developed, strain rate at the interface between the eddies enhances the mixing. During the eddy break-up process and the formation of smaller eddies, strain rate increases and thereby steepens the concentration gradients at the interface between reactants, which in turn enhances their molecular diffusion rate. Molecular mixing of fuel and oxidizer, a prerequisite of combustion, takes place at the interface between small eddies [4]. The subject of turbulent combustion spans a broad range of disciplines ranging from turbulent flows to combustion chemistry, which makes the analysis of turbulent combustion a daunting task. At the heart of the challenge is the presence of a broad range of length and time scales of the various processes governing combustion and the degree of coupling between these processes across all scales [5].

3.1 Turbulent scales

In order to estimate whether the chemistry is fast or slow compared to turbulent mixing, it is useful to define the time, length and velocity scales associated with physical processes. First consider the range of length scale (eddy sizes) that one may expect to encounter in turbulent flows. The largest length scale of turbulence is known as the integral length scale (L_{11}), which is the length scale at which most of the energetic eddies are associated. By contrast, the smallest length scale of turbulence known as Kolmogorov length scale, is determined by viscous dissipation of turbulent kinetic energy. From the Kolmogorov's hypothesis, the only factors influencing the behavior of the small scale motions are the overall kinetic energy dissipation rate (ε) and the viscosity (ν). The length scale which governs these physical mechanisms is given by:

$$\eta = \left(\frac{\nu^3}{\varepsilon}\right)^{1/4} \tag{1}$$

This length scale is called the Kolmogorov length scale (η) and is the smallest hydrodynamic scale in turbulent flows. The kinetic energy of the flow is proportional to (u'^2), where u' is the velocity fluctuation. The time scale of the life time of the large eddies (commonly referred to as the large eddy turnover time) can be estimated as L_{11}/u'. Therefore, the kinetic energy dissipation rate is:

$$\varepsilon \sim \frac{u'u'}{L_{11}/u'} \sim \frac{(u')^3}{L_{11}} \tag{2}$$

The ratio of the largest to smallest length scales in the turbulent flow is given by:

$$\frac{L_{11}}{\eta} \sim \left(\frac{u'L_{11}}{\nu}\right)^{3/4} \sim Re_t^{3/4} \tag{3}$$

where Re_t is the turbulent Reynolds number. Another commonly encountered length scale in turbulence is the Taylor micro-scale. This length scale does not have the same easily understood physical significance as the Kolmogorov or internal length scale but provides a convenient estimate for the fluctuating strain rate field. The Taylor micro-scale (λ) is defined through the relation:

$$\varepsilon \sim \nu \frac{\partial u_i'}{\partial x_j}\frac{\partial u_i'}{\partial x_j} \sim \frac{u'^2}{\lambda^2} \tag{4}$$

The large eddy turnover time (t_e) can be defined as $t_e \sim L_{11}/u'$. The life time for the small eddies of turbulence can be estimated using the viscosity and the dissipation rate as $t_\eta \sim \sqrt{\nu/\varepsilon}$. The ratio of time scales is therefore:

$$\frac{t_e}{t_\eta} \sim \left(\frac{u'L_{11}}{\nu}\right)^{1/2} \sim Re_t^{1/2} \tag{5}$$

The large scale structures in the flow are seen to have a much larger time scale (duration) than the smallest energy dissipating eddies. As the turbulent Reynolds number of the flow increases, the magnitude of the separation between both time and length scales increases. One can define computational time for DNS scales with Re_t^3.

3.2 Premixed turbulent combustion regimes

Diagrams defining regimes of premixed turbulent combustion in terms of velocity and length scale ratio have been proposed in a number of previous analyses [6–8]. For scaling purposes it is useful to assume equal diffusivities for all reactive scalars, Schmidt number $Sc = \nu/D$ equal to unity, where D is the mass diffusivity. Based on this one can define flame thickness as $l_f = D/S_b$ and flame time as

$t_f = D/S_b{}^2$, where S_b is the laminar burning velocity at given equivalence ratio. One can estimate the turbulent Reynolds number as:

$$Re_t \sim \frac{u' L_{11}}{S_b l_f} \tag{6}$$

Furthermore, one can quantify the separation between chemical time scale to the Kolmogorov time scale using Karlovitz number (Ka) as:

$$Ka \sim \frac{t_f}{t_\eta} \sim \frac{l_f^2}{\eta^2} \tag{7}$$

Figure 3 shows the the regime diagram for premixed turbulent combustion using the definition of Kolmogorov length scale. Moreover, **Figure 3** shows the typical working conditions realised in IC engines, gas turbines and counter flow regime on the regime diagram. Here the ratios u'/S_b and L_{11}/l_f may be expressed in terms of the two non-dimensional numbers Re_t and Ka as:

$$\frac{u'}{S_b} = Re_t \left(\frac{L_{11}}{l_f}\right)^{-1} = Ka^{2/3} \left(\frac{L_{11}}{l_f}\right)^{1/3} \tag{8}$$

The lines $Re_t = 1$ and $Ka = 1$ represents boundaries between different regimes of premixed turbulent combustion in **Figure 3**. Other boundaries of interest are the line $u'/S_b = 1$ which separated the wrinkled flamelets from the corrugated flamelets, and the line denoted by $Ka = 100$, which separates the thin reaction zones from broken reaction zones. The line $Re_t = 1$ separates all turbulent flame regimes characterised by $Re_t > 1$ from the laminar flame regime ($Re_t < 1$), which is situated in the lower-left corner of the diagram. In the wrinkled flamelet regime, where $u' < S_b$, the turnover velocity u' of even the integral eddies is not large enough to compete with the advancement of the flame front with the laminar burning velocity S_b. Laminar flame propagation therefore dominates over flame front corrugations by turbulence [8].

The corrugated flamelet regime is characterised by the inequalities $Re_t > 1$ and $Ka < 1$. The inequality indicates that $l_f < \eta$ (see Eq. (7)), which means that the entire

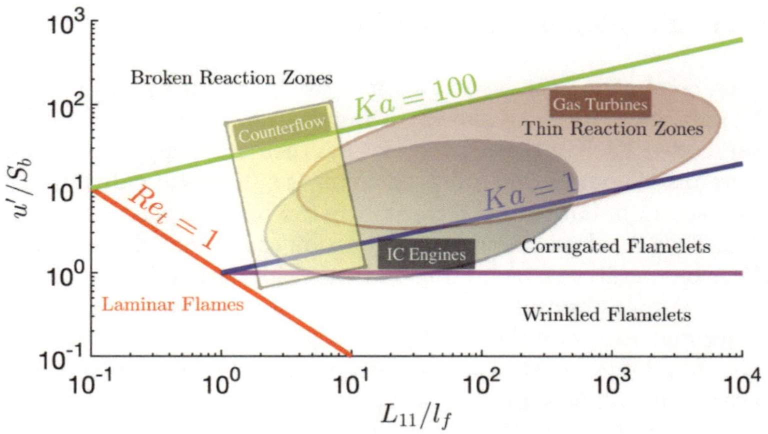

Figure 3.
Premixed turbulent combustion regime diagram accentuating areas for engineering applications.

Combustion regimes	Range
Laminar flames	$Re_t \leq 1$
Wrinkled flamelets	$Re_t \geq 1$; $u'/S_b \leq 1$
Corrugated flamelets	$Re_t \geq 1$; $u'/S_b > 1$; $Ka < 1$
Thin reaction zones[a]	$Re_t \geq 1$; $1 \leq Ka < 100$
Broken reaction zones	$Re_t \geq 1$; $Ka \geq 100$

[a]*The analyses argued that since quenching by vortices occurs only for large Karlovitz numbers, the region below the limiting value of the Karlovitz number should corresponds to the flamelet regime [8].*

Table 1.
Summary of premixed turbulent combustion regimes.

reactive-diffusive flame structure is embedded within the eddies of the size of the Kolmogorov scale, where the flow is quasi-laminar. Therefore the flame structure is not perturbed by turbulent fluctuation and retains its quasi-laminar structure [8].

The thin reaction zones regime is characterised by $Re_t > 1$ and $1 < Ka < 100$, the last inequality indicating that the smallest eddies of size η can enter into the reactive–diffusive flame structure since $\eta < l_f$ (see Eq. (7)). These small eddies are still larger than the reaction layer thickness l_δ and can therefore not penetrate into that layer. The thickness δ of the inner layer in a premixed flame is typically one tenth of the flame thickness, such that l_δ is one tenth of the preheat zone thickness which is of the same order of magnitude as the flame thickness l_f.

Beyond the line $Ka = 100$ there is a regime called the broken reaction zones regime where Kolmogorov eddies are smaller than the inner layer thickness l_δ. These eddies may therefore enter into the inner layer and perturb it with the consequence that chemical processes are disturbed locally owing to enhanced heat loss to the preheat zone followed by temperature decrease and the loss of radicals. When this happens the flame will extinguish and fuel and oxidizer will inter diffuse and mix at lower temperatures where combustion reaction has ceased to take place. Nevertheless, regime diagram provides a useful purpose in allowing the classification of turbulent premixed flames, different premixed turbulent combustion regimes are summarised in **Table 1**.

4. Combustion modes: premixed and non-premixed

Generally, combustion can be divided into two categories: premixed and non-premixed combustion. **Figure 4** shows a Venn diagram representing different combustion modes. Each of these categories has their advantages and disadvantages, but premixed combustion offers advantages in terms of pollutant emission because the maximum burned gas temperature can be controlled by the mixture composition. Thus fuel lean premixed combustion can potentially lead to reduction of burned gas temperature which offers reduction in thermal NOx emission [9]. In the demand to reduce harmful emissions, industrial combustors are designed to operate under fuel lean conditions and with inhomogeneous mixtures, which increasingly often leads to stratified combustion [10, 11]. Many engineering combustion systems including: lean premixed prevaporised (LPP) gas turbine combustor, afterburners, and direct-injection spark-ignition internal combustion engines, they all operate in inhomogeneous reactants mode to gain full advantages of a spatially varying mixture field [12–14].

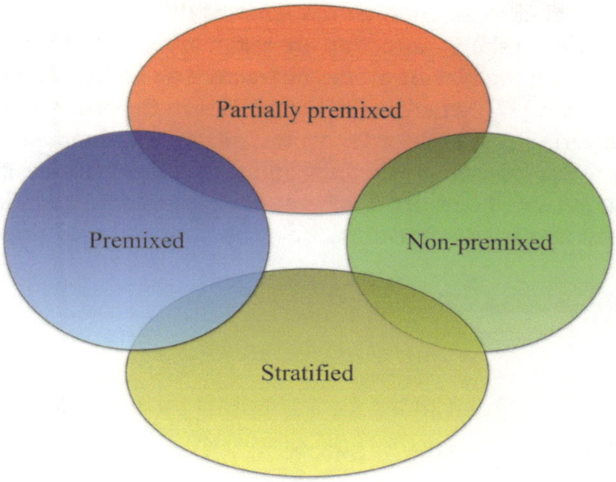

Figure 4.
Venn diagram representing combustion modes.

Stratified premixed combustion combines advantages of both premixed and non-premixed combustion modes (see **Figure 4**). In stratified combustion a premixed flame originated from ignition source travels through mixture field of varying equivalence ratio, which may be either all lean or all rich and the flame propagation is strongly affected by local gradient of mixture field [15]. For instant, in a gasoline direct injection (GDI) spark ignition engine, the time interval between fuel being injected into the combustion chamber and the spark ignition may be too short for the mixture composition to be homogeneous at the instant of ignition, but it is long enough for most of the fuel to be mixed with air before burning. The flame kernel originated by the spark propagates through a highly inhomogeneous mixture field characterised by large fluctuations in the equivalence ratio, with the ensemble-averaged mixture composition being lean (and even beyond lean flammability limit) in some spatial regions and rich (and even beyond the rich flammability limit) in other regions. In such example inhomogeneously premixed combustion is important, as it controls majority of the total heat release, while the afterburning of the lean and rich products in the diffusion mode may be of significant importance as far as pollutant (e.g. soot formation) is concerned. By contrast, in a diesel engine, the time interval between fuel injection and autoignition is too short that only a small amount of the fuel is mixed with air before autoignition of mixture due to compression. Here, also lean and rich premixed turbulent flames coexist with diffusion flames, but contrary to GDI engine, the total heat release is mainly controlled by the non-premixed mode of burning, and such regime is called nonpremixed/premixed combustion [7, 16]. All aforementioned combustion modes (stratified, premixed/nonpremixed, and nonpremixed/premixed burning) are commonly absorbed under partially-premixed flames [7].

5. DNS results and discussions

This section includes some of the DNS results of turbulent combustion in different environments. The results are presented and subsequently discussed. All the simulations presented here are performed using a well known compressible DNS code SENGA [3]. This DNS code solves the full compressible Navier-Stokes equations on a cartesian grid. The governing equations that describe the 3D gaseous reacting flow consists of mass, momentum, energy and species conservation equations. The boundaries in the x_1–direction are taken to be partially non-reflecting and are specified using the Navier-Stokes characteristic boundary conditions (NSCBC) formulation [17], whereas the boundaries in the other directions are considered to be periodic. A 10th-order central difference scheme is used for spatial differentiation for the internal grid points, and the order of differentiation gradually reduces to a one-sided 2nd-order scheme at non-periodic boundaries [18]. The time advancement is carried out using a 3rd-order low-storage Runge–Kutta scheme [19].

5.1 Ignition

A source term q''' has been added for accounting the heat addition to the energy transport equation:

$$\frac{\partial \rho E}{\partial t} + \frac{\partial \rho u_k E}{\partial x_k} = -\frac{\partial u_k P}{\partial x_k} + \frac{\partial \tau_{ki} u_i}{\partial x_k} + \frac{\partial}{\partial x_k}\left[\lambda \frac{\partial \hat{T}}{\partial x_k}\right] + \dot{w}_T + q''' \tag{9}$$

The specific heats at constant pressure and constant volume (i.e. c_p and c_v) are taken to be constant and same for all species for the simplified chemistry results. Therefore, the term $\sum_{k=1}^{N} h_{s,k} Y_k V_{k,i} = 0$ and not included in the Eq. (9). The source term is assumes to follow a Gaussian distribution [20] in the radial direction away from the centre of the igniter and can be expressed as:

$$q'''(r) = A_q \exp\left(-\frac{r^2}{2R^2}\right) \tag{10}$$

where r is the radial distance from the centre of the igniter and R is the width of the Gaussian profile. The choice of R [21] in the present analysis allows sufficient resolution of the temperature gradient and guarantees the rapid disappearance of any artificial effects introduced by the ignition source. The constant A_q is determined by the following volumetric integration [21]:

$$\dot{Q} = \int_V q''' dV \tag{11}$$

where $\dot{Q}$ is the ignition power, which can be defined as:

$$\dot{Q} = a_{sp} \rho_0 c_p \tau T_0 \left(\frac{4}{3}\pi l_f^3\right)\left[\frac{H(t) - H(t - t_{sp})}{t_{sp}}\right] \tag{12}$$

where a_{sp} is a parameter that determined the total energy deposited by the igniter, the t_{sp} is the time duration over which the energy is deposited by the igniter, which is expressed as $t_{sp} = b_{sp} \times t_f$, where b_{sp} is the energy duration parameter and t_f is a characteristic chemical time scale given by $t_f = l_f / s_b$. The parameter b_{sp} varies between 0.2 and 0.4 for optimum spark duration according to experimental findings [22]. Here l_f is the Zel-dovich flame thickness for the stoichiometric mixture, which is defines as $l_f = \alpha T_0 / S_b$, where αT_0 is the thermal diffusivity in the unburned reactants and S_b is the unstrained laminar burning velocity of the stoichiometric mixture. It is important to note that the details of the spark formation (momentum modification contribution, plasma formation and shock wave) remain beyond the scope of the present analysis to keep this study computationally feasible. Parametric study of the effects of energy deposition characteristics (a_{sp} [parameter determining total energy deposition], b_{sp} [energy duration parameter] and R [parameter determining characteristic width of ignition energy deposition]) on localised forced ignition and early stages of burning following successful ignition in homogeneous turbulent mixtures using 3D DNS are investigated by [23, 24].

5.2 Ignition in homogeneous mixture

5.2.1 Isosurface of temperature field

Premixed combustion can be described in terms of a composition variable known as reaction progress variable. This reaction progress variable describe the progress of the premixed reaction [25]. The active scalars which are often considered for analysing turbulent combustion are the fuel mass fraction Y_F and the reaction progress variable c [26]. The extent of the completion of chemical reaction can be quantified in terms of a reaction progress variable c, defined as:

$$c = \frac{Y_{Fu} - Y_F}{Y_{Fu} - Y_{Fb}} \tag{13}$$

where Y_{Fu} is the fuel mass fractions in the unburned gas and Y_{Fb} is the fuel mass fraction in the burned gas. Both Y_{Fu} and Y_{Fb} are the function of mixture fraction. According to Eq. (13), c rises monotonically from zero in the fully unburned reactants to unity in the fully burnt products. **Figure 5** shows 3D volume rendered view of non-dimensional temperature for different Karlovitz numbers ranging across different turbulent regimes. The values for non-dimensional temperature T corresponds to reaction progress variable $c \geq 0.9$ showing the flame kernel. It is evident from **Figure 5** that the isosurfaces of T remains approximately spherical during the period of energy deposition, but they become increasingly wrinkled as time progresses for all the cases. During the energy deposition period (i.e. $t \leq t_{sp}$) the evolution of temperature field is principally determined by the diffusion of energy deposited, but after the ignition initiated, the flame propagation controlled by the diffusion of local flame stretching mechanism. The turbulence tends to fragment the flame surface and breaks the reaction zone for higher Karlovitz number cases. Furthermore interesting observation is when $t \gg t_{sp}$, the high temperature region has been fragmented thanks to energetic eddies of turbulence penetrating into the flame. This tendency is more prominent for $Ka = 150$, where turbulent eddies enters into flame and start breaking the reaction zone and flame eventually extinguishes.

5.2.2 Reaction-diffusion balance analysis

It is extremely important that heat release due to chemical reaction should overcome the heat transfer from the hot gas kernel in order to obtain self-sustained combustion following successful ignition. The transport equation of the reaction progress variable c in the context of turbulent premixed flames is:

$$\frac{\partial \rho c}{\partial t} + \frac{\partial \rho u_j c}{\partial x_j} = \frac{\partial}{\partial x_j}\left[\frac{\partial \rho D c}{\partial x_j}\right] + \dot{w}_c \tag{14}$$

where $\dot{w}_c$ is the reaction progress variable reaction rate.

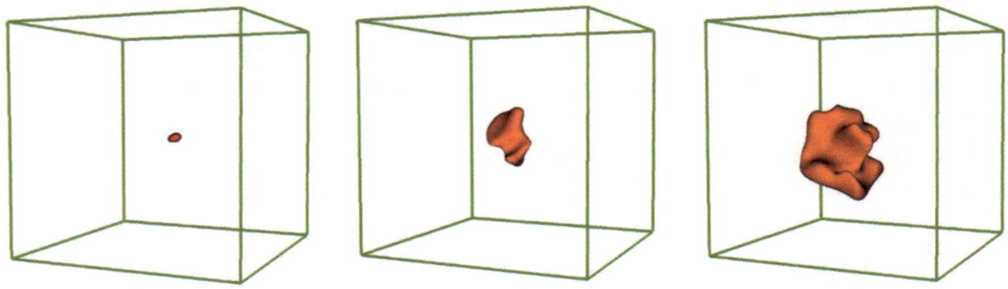

(a) $Ka = 6.30$ at $2.1t_{sp}$, $6.4t_{sp}$ and $12.6t_{sp}$

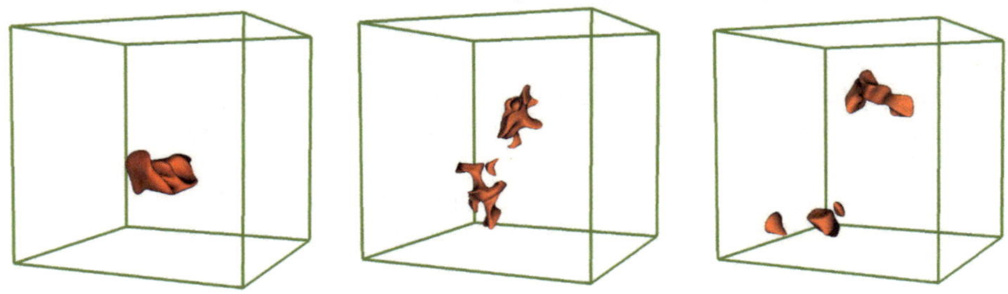

(b) $Ka = 44.20$ at $2.1t_{sp}$, $6.4t_{sp}$ and $12.6t_{sp}$

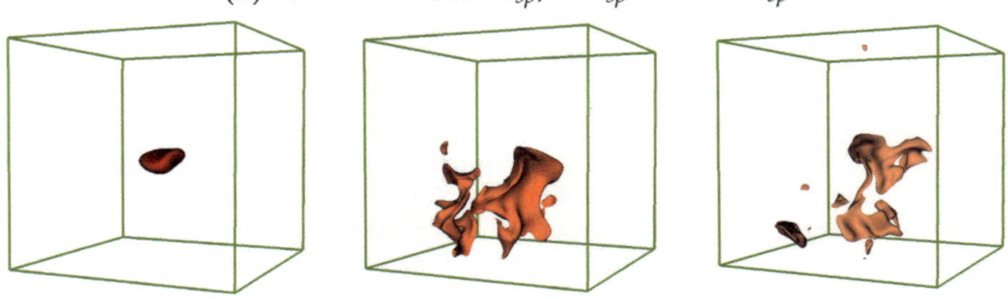

(c) $Ka = 150$ at $2.1t_{sp}$, $6.4t_{sp}$ and $12.6t_{sp}$

Figure 5.
Volume rendered view of non-dimensional temperature field in domain $51l_f \times 51l_f \times 51l_f$ for different Karlovitz numbers (Ka) in different turbulent regimes.

It is important to examine the reaction–diffusion balance in order to understand the difference in flame kernels which are growing in an unperturbed manner and those which are fragmented and about to be quenched by turbulence. In that respect, following terms are defined to explain the reaction–diffusion balance within the flame:

$$\text{reaction term}: \quad \dot{w}_c \tag{15}$$

$$\text{molecular diffusion term}: \quad \nabla \cdot (\rho D \nabla c) \tag{16}$$

$$\text{flame normal diffusion rate term}: \quad \vec{N} \cdot \nabla \left(\rho D \vec{N} \cdot \nabla c \right) \tag{17}$$

$$\text{tangential diffusion rate term}: \quad -\rho D \nabla \cdot \vec{N} |\nabla c| \tag{18}$$

where $\vec{N}$ is the flame normal vector, which can be defined as:

$$\vec{N} = -\frac{\nabla c}{|\nabla c|} \tag{19}$$

It can be seen from **Figure 6** that the term $\dot{w}_c$ (■) remains negligible in the unburned side and increases sharply towards the burned side before decreasing to zero in the fully burned products. The magnitude of this term decreases with time, once the igniter is switched off, which is principally due to the decrease in fuel reaction rate magnitude with time. Therefore, it can be seen that maximum reaction takes place in the region $0.6 \leq c \leq 0.99$. The term $\nabla \cdot (\rho D \nabla c)$ (■) shows opposite behaviour to the term $\dot{w}_c$ (■), suggesting reactants are diffusing prior to approaching the flame kernel and term remains negative in burned side. Additionally, the term $\dot{w}_c + \nabla \cdot (\rho D \nabla c)$ (■) suggests a self-sustained propagation of flame kernel. The term $\vec{N} \cdot \nabla\left(\rho D \vec{N} \cdot \nabla c\right)$ (■) remains positive in unburned side but becomes negative towards burned side. The term $-\rho D \nabla \cdot \vec{N} |\nabla c|$ (■) remains negative for the flame brush. In the flame normal vector, N_j is the jth component of flame normal which can be defined in terms of flame curvature [27]:

$$\kappa_m = \frac{1}{2} \nabla \cdot \vec{N} = \frac{\kappa_1 + \kappa_2}{2} \sim \frac{1}{R_{fk}} \tag{20}$$

where κ_1 and κ_2 are the two principle curvature of the concerned flame surface, κ_m is the arithmetic mean of these two principle curvatures and R_{fk} is the radius of a flame kernel. As the flame kernel increases in size (i.e. R_{fk} increases), the probability of high values of κ_m decreases (see Eq. (20)). Furthermore according to Eq. (20) positively curved location (i.e. $\kappa_m > 0$) are convex towards the reactant. The flame is initiated as a spherical kernel from the localised forced ignition, which has a positive mean curvature. In this respect the term $-\rho D \nabla \cdot \vec{N} |\nabla c|$ (■) will attain negative values in the case of expanding flame kernel (i.e. self-sustained propagating flame kernel). The growth of the kernel leads to a decrease in the magnitude of the negative contribution of the term $-\rho D \nabla \cdot \vec{N} |\nabla c|$ (■), whereas the term $-\rho D \nabla \cdot \vec{N} |\nabla c|$ (■) is expected to assume large negative values in the kernels which are quenching. This findings are in good agreement with previous experimental studies [28, 29].

5.3 Ignition in stratified mixture

Premixed combustion offers an option of controlling flame temperature and reducing pollutant emission such as nitrogen oxides (NOx) but, in practice, perfect mixing is often difficult to achieve and thus combustion in many engineering applications takes place in turbulent stratified mixtures. Many previous findings

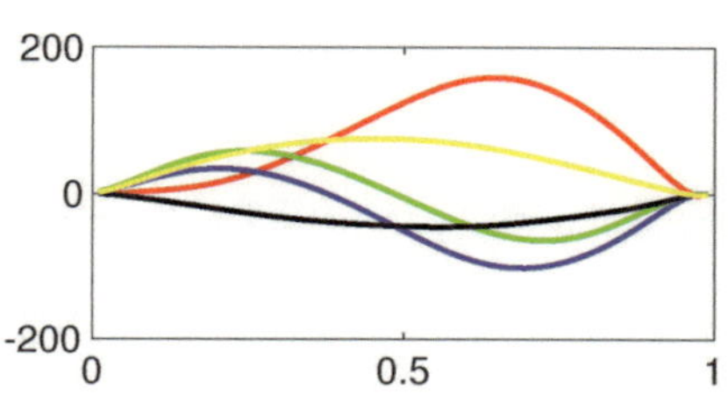

$\dot{w}_c$ (■); $\nabla \cdot (\rho D \nabla c)$ (■); $\vec{N} \cdot \nabla(\rho D \vec{N} \cdot \nabla c)$ (■); $-\rho D \nabla \cdot \vec{N} |\nabla c|$ (■); $\dot{w}_c + \nabla \cdot (\rho D \nabla c)$ (■);

Figure 6.
Mean variation of different terms with reaction progress variable at $t = 2.6 t_{sp}$ for $Ka = 6.30$.

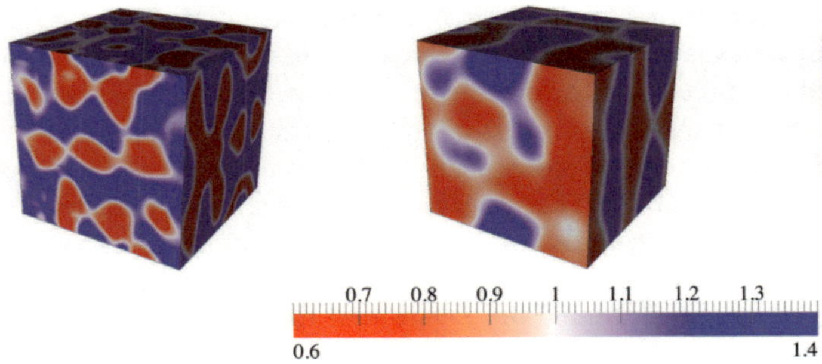

Figure 7.
Initial computational domain showing different values of length scale of mixture inhomogeneities, l_ϕ/l_f = 2.1 *(left),* =5.5 *(middle), and* =8.3 *(right).*

[30, 31] shows that the flame propagation statistics are strongly influenced by the local equivalence ratio gradient. The length scale of mixture inhomogeneity is taken as the Taylor micro-scale of the equivalence ratio variation l_ϕ and is defined as [32]:

$$l_\phi = \sqrt{\frac{6\left\langle [\phi - \langle\phi\rangle]^2 \right\rangle}{\langle \nabla[\phi - \langle\phi\rangle] \cdot \nabla[\phi - \langle\phi\rangle] \rangle}} \tag{21}$$

The equivalence ratio ϕ variation is initialised using a random distribution of ϕ following a Bi-modal distribution for specified values of the mean global equivalence ratio $\langle\phi\rangle$. In practical, when fuel is introduced in the liquid phase, the probability density function (PDF) of the equivalence ratio distribution is likely to be Bi-modal as a result of localised liquid fuel evaporation during the early stage of mixing. The fuel-air mixture is likely to be fuel rich close to the evaporation sites and the mixture is expected to be fuel-lean far away from the droplets. The initial mixture distribution for $\langle\phi\rangle = 1$ and ϕ' = 0.4 with different values of l_ϕ/l_f are shown in **Figure 7**, which indicated that the clouds of mixture inhomogeneities increase in size with increasing l_ϕ/l_f.

5.3.1 Mode of combustion

The role of the reaction progress variable c, in the turbulent stratified mixtures has been discussed in detail by Bray et al. [33]. In the context of stratified combustion, the reaction progress variable can be defines in the following manner [21, 25]:

$$c = \frac{\xi Y_{F\infty} - Y_F}{\xi Y_{F\infty} - \max\left[0, \frac{\xi - \xi_{st}}{1 - \xi_{st}}\right] Y_{F\infty}} \tag{22}$$

where ξ is the conserved scalar and can be defined as:

$$\xi = \frac{Y_F - \frac{Y_O}{s} + \frac{Y_{O\infty}}{s}}{Y_{F\infty} + \frac{Y_{O\infty}}{s}} \tag{23}$$

where s is the mass of the oxidiser consumed per unit mass of fuel consumption, Y_F and Y_O are local fuel and oxidiser mass fractions respectively, $Y_{F\infty}$ is the fuel mass fraction in pure fuel stream and $Y_{O\infty}$ is the oxidiser mass fraction in air.

In order to understand the flame structure originating from localised forced ignition (e.g. spark or laser), the Takeno flame index [34, 35] can be used to identify the local combustion mode:

$$I_c = \frac{\nabla Y_F}{|\nabla Y_F|} \cdot \frac{\nabla Y_O}{|\nabla Y_O|} \tag{24}$$

Based on Eq. (24), the Takeno flame index obtains positive value in premixed mode of combustion and negative value in non-premixed mode of combustion. The volume rebdered views of the region corresponding to $0.01 \le c \le 0.99$ coloured with local values of I_c at $t = 8.40t_{sp}$ are shown in **Figure 8** for selected cases. It is evident from **Figure 8** that the reaction takes place predominantly in the premixed mode (i.e. $I_c > 0$) but some pockets of $I_c < 0$ indicate the possibility of finding local pockets of non-premixed combustion. The probability of finding $I_c < 0$ decreases with increasing time due to mixing process. The I_c predominantly assumes positive values and major portion of overall heat release arises due to the premixed mode of combustion in all cases. Moreover, the percentage of heat release arising from the non-premixed mode of combustion decreases with increasing value of u' as a result of improved mixing.

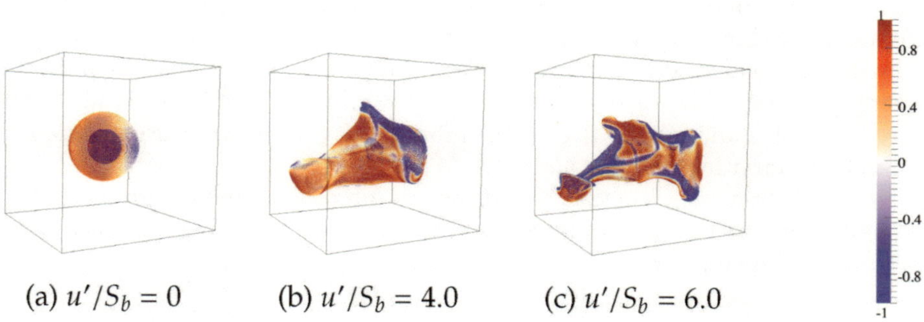

(a) $u'/S_b = 0$ (b) $u'/S_b = 4.0$ (c) $u'/S_b = 6.0$

$I_c > 0$ (■); $I_c < 0$ (■)

Figure 8.
Volumetric rendering of the region corresponding to $0.01 \le c \le 0.99$ in domain $33l_f \times 33l_f \times 33l_f$ coloured with local values of I_c for different values of u'/S_b and mixture inhomogeneity $l_\phi/l_f = 8.3$.

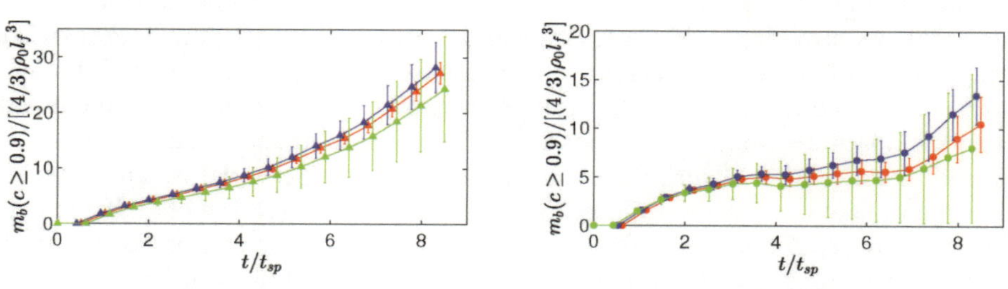

(a) $u'/S_b = 0$ (b) $u'/S_b = 4.0$

$l_\phi/l_f = 2.1$ (■); $l_\phi/l_f = 5.5$ (■); $l_\phi/l_f = 8.3$ (■)

Figure 9.
Temporal evolution of mean burnt gas mass with $\phi' = 0.2$ and $l_\phi/l_f = 5.5$ with standard deviation due to different realisation of initial conditions shown in the form of bars.

5.3.2 Extent of burning

The extent of burning can be characterised by the mass of burned gas m_b with $c \geq 0.9$ [21]. The temporal evolution of burnt gas mass normalised by the mass of an unburned gas sphere with a radius equal to l_f for $l_\phi/l_f = 5.5$ are show in **Figure 9**. It is important to note that $\int_V \rho c \, dV$ provides the measure of total burned gas mass within the computational domain. **Figure 9** shows that an increase in u' leads to reduction in burned gas mass for all cases. An increase in u' leads to an increase in eddy diffusivity $D_t \sim u' L_{11}$ for a given high values of turbulence intensity. For self-sustained flame propagation following successful ignition, the heat release from the combustion must overcome the heat loss from hot gas kernel. Due to heat transfer from hot gas kernel, the probability of $c \geq 0.8$ also decreases and at such point the heat loss overcomes chemical heat release, the hot gas kernel shrinks and eventually leads to flame extinction. The detrimental effect of u' on the extent of burning is consistent with previous findings [36]. The influence of l_ϕ/l_f on the extent of burning is found to be non-monotonic and dependent on ϕ'. For high values of l_ϕ/l_f the clouds of mixture stratification are relatively big and as a result of this, there is a possibility that igniter has encountered a rich region (highly flammable) of mixture which leads to higher burning rate. It is well known that combustion succeeds only for selected realisation of initial distribution of ϕ and this aspect is particular interest in the cylinder of IC engines due to cycle-to-cyle variation [20]. The temporal evolution of mean and standard deviations of burnt gas mass for all realisations are shown in **Figure 9** to demonstrate the probabilistic nature of the localised ignition of stratified mixtures. Furthermore, the large variation of error bar for hight values of l_ϕ/l_f suggests that it is possible to obtain large clouds of both highly flammable and weakly flammable mixtures at the igniter location which leads to large variation of burnt gas mass.

The above discussion suggests that turbulent intensity u' and length scale of mixture inhomogeneity l_ϕ have important influences on achieving self-sustained combustion following successful ignition event.

6. Epilogue

Theories and results presented in this chapter suggests that turbulent premixed combustion is a complex and difficult subject, but very rich in the physics. With recent advances in computational capability, the application of DNS will become possible for higher values of turbulent Reynolds number and complex flow configurations. DNS provides highly accurate and detailed 3D information in comparison to experiments because it is extremely expensive or impossible to obtain 3D temporally and spatially resolved data by experimental means. However, with current advances in laser technology, it is possible to have simultaneous planer laser-induced measurement of turbulent concentration and velocity fields. Once this experimental data becomes available, it will be used for validation of DNS results. Moreover, successful ignition often leads to momentum modification contribution, plasma formation, and shock waves, which remains beyond the scope of the present DNS analysis. Additionally, detailed chemical mechanism involving large number of intermediate species which lead to back-diffusion of light radical and post diffusion flames are necessary to gain further fundamental understanding of turbulent combustion processes.

Acknowledgements

Authors are grateful to facilities of Compute Canada for computational support. The author gratefully acknowledge the training he receive from his supervisor Prof. N. Chakraborty. The author also acknowledge the practical help and valuable discussion with Dr. Jiawei Lai, Dr. Sahin Yigit and Dr. Bruno Machado while preparing the manuscript.

Conflict of interest

The authors declare no conflict of interest.

Author details

Dipal Patel† and Martin Agelin-Chaab*†
Department of Automotive, Mechanical and Manufacturing Engineering, Faculty of Engineering and Applied Science, Ontario Tech University, Oshawa, ON, Canada

*Address all correspondence to: martin.agelin-chaab@uoit.ca

† These authors contributed equally.

References

[1] Hirsch C. Numerical and Computation of Internal and External Flows: The Fundamentals of Computational Fluid Dynamics. Elsevier; 2007

[2] Roache PJ. Fundamentals of Computational Fluid Dynamics. Albuquerque, NM: Hermosa Publishers; 1998

[3] Cant S. Direct numerical simulation of premixed turbulent flames. Philosophical Transactions of the Royal Society of London. Series A: Mathematical, Physical and Engineering Sciences. 1999;**357**(1764):3583-3604

[4] Poinsot T, Veynante D. Theoretical and Numerical Combustion. RT Edwards, Inc.; 2005. ISBN 1930217102

[5] Echekki T, Mastorakos E. Turbulent Combustion Modeling: Advances, New Trends and Perspectives, Volume 95. Springer Science & Business Media; 2010. ISBN: 9400704127

[6] Bradley D, Lung FK-K. Spark ignition and the early stages of turbulent flame propagation. Combustion and Flame. 1987;**69**(1):71-93

[7] Lipatnikov A. Fundamentals of Premixed Turbulent Combustion. CRC Press; 2012. ISBN: 1466510250

[8] Peters N. Turbulent Combustion. Cambridge University Press; 2000

[9] McAllister S, Chen J-Y, Carlos Fernandez-Pello A. Fundamentals of Combustion Processes. Vol. 302. New York: Springer; 2011

[10] Alkidas AC. Combustion advancements in gasoline engines. Energy Conversion and Management. 2007;**48**(11):2751-2761

[11] Drake MC, Haworth DC. Advanced gasoline engine development using optical diagnostics and numerical modeling. Proceedings of the Combustion Institute. 2007;**31**(1):99-124

[12] Brosh T, Patel D, Wacks D, Chakraborty N. Numerical investigation of localized forced ignition of pulverized coal particle-laden mixtures: A Direct Numerical Simulation (DNS) analysis. Fuel. 2015;**145**:50-62

[13] Lim KM. DNS of inhomogeneous reactants premixed combustion [PhD thesis]. Fitzwilliam College; 2015

[14] Mongia HC, Held TJ, Hsiao GC, Pandalai RP. Challenges and progress in controlling dynamics in gas turbine combustors. Journal of Propulsion and Power. 2003;**19**(5):822-829

[15] Swaminathan N, Grout R, Mastorakos E. Direct simulations of forced ignition in stratified mixture. In: Proceedings of the Third European Combustion Meeting. 2007

[16] Bilger RW, Pope SB, Bray KNC, Driscoll JF. Paradigms in turbulent combustion research. Proceedings of the Combustion Institute. 2005;**30**(1):21-42

[17] Poinsot TJ, Lelef SK. Boundary conditions for direct simulations of compressible viscous flows. Journal of Computational Physics. 1992;**101**(1): 104-129

[18] Jenkins KW, Cant RS. Flame kernel interactions in a turbulent environment. In: Technical Report. Cambridge University; 2001

[19] Wray AA. Minimal storage time advancement schemes for spectral methods. In: Report No. MS 202. California: NASA Ames Research Center; 1990

[20] Pera C, Chevillard S, Reveillon J. Effects of residual burnt gas

heterogeneity on early flame propagation and on cyclic variability in spark-ignited engines. Combustion and Flame. 2013;**160**(6):1020-1032

[21] Patel D, Chakraborty N. Localised forced ignition of globally stoichiometric stratified mixtures: A numerical investigation. Combustion Theory and Modelling. 2014;**18**(6):627-651

[22] Ballal D, Lefebvre A. Spark ignition of turbulent flowing gases. Aerospace Research Central. 1977:185

[23] Patel D, Chakraborty N. Effects of energy deposition characteristics on localised forced ignition of homogeneous mixtures. International Journal of Spray and Combustion Dynamics. 2015;**7**(2):151-174

[24] Patel D, Chakraborty N. Effects of fuel Lewis number on localised forced ignition of turbulent homogeneous mixtures: A numerical investigation. International Journal of Spray and Combustion Dynamics. 2016;**8**(3):183-196

[25] J Hélie and A Trouvé. Turbulent flame propagation in partially premixed combustion. In Turbulent Flame Propagation in Partially Premixed Combustion, volume 27, pages 891–898. Elsevier, 1998. ISBN 0082-0784

[26] Malkeson SP, Chakraborty N. Statistical analysis of displacement speed in turbulent stratified flames: A direct numerical simulation study. Combustion Science and Technology. 2010;**182**(11–12):1841-1883

[27] Chakraborty N, Mastorakos E. Numerical investigation of edge flame propagation characteristics in turbulent mixing layers. Physics of Fluids. 2006; **18**(10):105103

[28] Ahmed SF, Mastorakos E. Spark ignition of lifted turbulent jet flames. Combustion and Flame. 2006;**146**(1–2): 215-231

[29] Chung SH. Stabilization, propagation and instability of tribrachial triple flames. Proceedings of the Combustion Institute. 2007;**31**(1): 877-892

[30] Haworth DC, Blint RJ, Cuenot B, Poinsot TJ. Numerical simulation of turbulent propane–air combustion with nonhomogeneous reactants. Combustion and Flame. 2000;**121**(3): 395-417

[31] Anselmo-Filho P, Hochgreb S, Barlow RS, Cant RS. Experimental measurements of geometric properties of turbulent stratified flames. Proceedings of the Combustion Institute. 2009;**32**(2):1763-1770

[32] Eswaran V, Pope SB. Direct numerical simulations of the turbulent mixing of a passive scalar. The Physics of fluids. 1988;**31**(3):506-520

[33] Bray K, Domingo P, Vervisch L. Role of the progress variable in models for partially premixed turbulent combustion. Combustion and Flame. 2005;**141**(4):431-437

[34] Hannebique G, Sierra P, Riber E, Cuenot B. Large eddy simulation of reactive two-phase flow in an aeronautical multipoint burner. Flow, Turbulence and Combustion. 2013; **90**(2):449-469

[35] Lignell D, Chen JH, Lu T, Law CK. Direct numerical simulation of extinction and reignition in a nonpremixed turbulent ethylene jet flame. In: Technical Report. Sandia National Lab.; 2007

[36] Richardson ES, Mastorakos E. Numerical investigation of forced ignition in laminar counterflow non-premixed methane-air flames. Combustion Science and Technology. 2007;**179**(1–2):21-37

Chapter 7

A Theoretical Review of Rotating Detonation Engines

Ian J. Shaw, Jordan A.C. Kildare, Michael J. Evans, Alfonso Chinnici, Ciaran A.M. Sparks, Shekh N.H. Rubaiyat, Rey C. Chin and Paul R. Medwell

Abstract

Rotating detonation engines are a novel device for generating thrust from combustion, in a highly efficient, yet mechanically simple form. This chapter presents a detailed literature review of rotating detonation engines. Particular focus is placed on the theoretical aspects and the fundamental operating principles of these engines. The review covers both experimental and computational studies, in order to identify gaps in current understanding. This will allow the identification of future work that is required to further develop rotating detonation engines.

Keywords: rotating detonation engine, detonative engines, propulsion, detonation shock waves, spin detonation engine

1. Introduction

1.1 Background

Detonative combustion is a potential propulsion method for aerospace systems, offering high efficiency and low mechanical complexity. In comparison, deflagration is generally considered easier to control and has therefore dominated both experimental and real world engine applications. Research into detonation engines has been limited due to the lack of the necessary tools required to design and analyse such systems [1, 2]. As such, practical development of detonation engines, notably the pulsed detonation engine (PDE) and the rotating or rotational detonation engine (RDE), has been limited [3]. Nevertheless, the application of detonation engines for propulsion is very promising, already proving to be compact, whilst providing highly efficient thrust generation [3–7]. This supersonic thrust could be utilised independently as a rocket engine, or as part of a gas turbine system. Interest in the development of RDE technology has grown and the challenges of utilising a more thermodynamically-efficient cycle have become better understood [8, 9].

Combustion can occur at both subsonic and supersonic velocities, known as deflagration and detonation, respectively. Deflagration is typified by a regular flame, which propagates at less than the speed of sound. The heat release may be used to expel the resulting products, generating thrust. Deflagration has been used in a broad range of applications to produce power. However, in theory, deflagration lacks the thermodynamic efficiency of a detonation system, which is a system where combustion is initiated suddenly and "propagates utilising most, if not all, of the heat from combustion in an incredibly rapid shock wave" [10]. The heat generated by the exothermic chemical reaction sustains the shock wave. The concept of using detonation as a propulsion source has been proposed since the 1840s [11], but no substantial work had been completed until the 1950s when the development of models and concepts for a more lightweight and compact engine began [12]. The mechanisms that drive the detonation engine were not well understood at that time, so much of the research over the following decades was centred on the theoretical development of the engine.

As the name implies, the pulse detonation engine (PDE) has been proposed for propulsion using detonations [12, 13]. In a PDE, a detonation chamber is filled with a fuel/oxidiser mixture, which is subsequently detonated. The accelerating detonation propels the exhaust from the chamber, thereby generating thrust. The chamber is then re-primed with fresh reactants, and re-detonated. With sufficiently high cycle speeds, large amounts of thrust may be generated in a small engine [14, 15]. This type of engine has been found to be particularly efficient [3, 16, 17].

Development of the concept of a rotating detonation engine (RDE) began as a result of further work into detonative propulsion. This engine type is characterised by one or more detonation waves contained within an open-ended annular chamber. A fuel/oxidiser mixture is fed into one end of the chamber, and the detonation wave consumes these reactants azimuthally, expelling reactants from the open end of the annulus. In some literature, this type of engine may also be referred to as a continuous detonation wave engine (CDWE) or a spin detonation engine [6].

Early research into rotating detonations was conducted in the 1950s [18], with attempts to document the structure of detonation shock waves, including those in spinning detonations, with further developments through the 1960s [1]. Subsequent research has been conducted into the effects of geometry, rotation characteristics, spiralling of the wave, and other variables [6, 19–22]. Another advancement in general detonation research is improvements in deflagration to detonation transitions (DDTs), leading to a greater understanding of the consumption of fuel in the chamber [23–25]. Further work has developed prototype RDEs to measure the thrust of small-scale units as a baseline for larger model behaviour, utilising the results from experimental work to verify theoretical results, and to generate new results [26–30].

In this review, several aspects of RDEs will be examined, starting with a brief comparison of RDEs and PDEs. This will be followed by further exploration into RDE operation, and methods of analysing RDEs, both experimentally and with numerical modelling. Finally, there will be an overview of areas still requiring further work.

1.2 Thermodynamic cycles

The majority of gas turbines that operate with a deflagration follow the Brayton (B) cycle: an isobaric (constant pressure) process, as shown in **Figure 1** [31]. In contrast, a detonation is almost isochoric (constant volume) and may be modelled with the Humphrey (H) cycle, or, preferably, with the Fickett-Jacobs (FJ) cycle, which models detonation [3, 31]. The H cycle assumes that combustion occurs in a fixed volume, resulting in a pressure spike as the products expand. Differentiation between the H and FJ cycles in **Figure 1** can be seen through the state changes of $2–3'$ for the H cycle and $2–3''$ for the FJ [31]. This pressure spike decreases the volume of combustion for FJ while remaining constant for H. The next phase (FJ $3''–4''$, H $3'–4'$) is similar for the two cycles, with the FJ cycle expanding further before reaching atmospheric pressure. Both then undergo a constant pressure compression through cooling back to the initial state 1. As seen in **Figure 1**, the FJ cycle is more

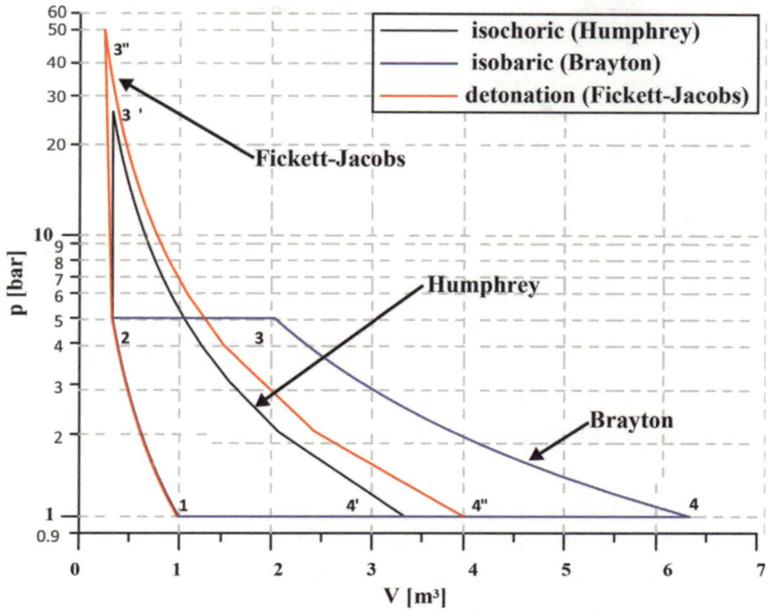

Figure 1.
Thermodynamic cycles: Humphrey, Brayton, and Fickett-Jacobs. Adapted from Wolański [31].

volumetrically efficient than the B cycle, and involves a higher pressure gain than the H, indicating that for the same initial isochoric compression, the FJ cycle is the more efficient of the three. This is supported by the thermodynamic efficiency equations for each of the cycles [31]:

$$\eta_B = 1 - \frac{1}{\left(\frac{p_2}{p_1}\right)^{\frac{k-1}{k}}} \tag{1}$$

$$\eta_H = 1 - k\frac{T_1}{T_2}\frac{\left(\frac{T_{3'}}{T_2}\right)^{\frac{1}{k}}}{\frac{T_{3'}}{T_2} - 1} \tag{2}$$

$$\eta_F = 1 - k\frac{1}{\left(\frac{p_3}{p_1}\right)^{\frac{k-1}{k}}}\frac{\left(\frac{T_{3''}}{T_2}\right)^{\frac{1}{k}} - 1}{\frac{T_{3''}}{T_2} - 1} \tag{3}$$

where η_B, η_H, and η_F are the thermal efficiencies of the Brayton, Humphrey, and Fickett-Jacobs cycles, T is temperature, p is pressure, k is the ratio of specific heats, and the numerical subscripts denote the position on the plot in **Figure 1** [31]. A substitution of the relevant temperatures, pressures, and specific heat ratios into the above equations indicate the higher thermal efficiency of the FJ cycle. Additionally, the thermal efficiencies of various fuels under each of these thermodynamic cycles have been calculated and reported in **Table 1**, further supporting the use of the FJ cycle when exploring detonation cycles as a high efficiency combustion method.

1.3 Pulsed detonation engines

In a PDE, such as that shown in **Figure 2**, a detonation chamber is filled with a fuel/oxidiser mixture and then ignited. The deflagration of the reactants accelerates,

Fuel	Brayton (%)	Humphrey (%)	Fickett-Jacobs (%)
Hydrogen (H_2)	36.9	54.3	59.3
Methane (CH_4)	31.4	50.5	53.2
Acetylene (C_2H_2)	36.9	54.1	61.4

Table 1.
Calculated thermodynamic efficiencies for various fuels under different thermodynamic cycles [26].

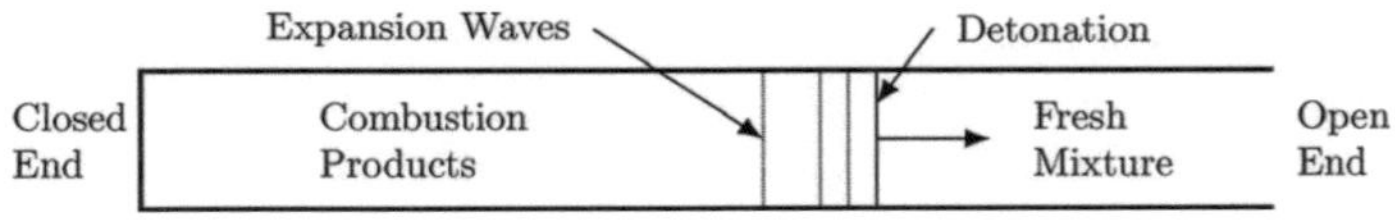

Figure 2.
Labelled schematic of a PDE. Adapted from [15].

and through a deflagration-to-detonation transition (DDT), generates a shock wave. The products are accelerated from the end of the chamber, carried by the detonation front, generating thrust [30, 31]. For each cycle, the chamber must be purged and then refilled with fresh fuel/oxidiser mixture and then detonated again, limiting the maximum practical frequency of operation to an order of 100 Hz [32]. This results in poor efficiency when scaled to high thrust levels as the discontinuous thrust cycles may not be fast enough to approximate the continuity required for propulsion purposes [32–35]. In some designs, it is also necessary to purge the chamber with an inert gas due to some residual combustion products remaining stagnant in the detonation chamber that interfere with the next detonation cycle. This process further restricts the operating frequency to approximately 50 Hz [3, 16].

In order to provide a more compact device, obstacles may be placed in the chamber to accelerate the DDT, but these reduce the specific impulse (I_{sp}) [31, 33]. Specific impulse can be defined as the change in momentum per unit mass of propellant used. An alternative approach is to remove the requirement for repeated DDT transitions, and hence the efficiency loss, by sustaining the detonation reaction. This approach leads directly to the concept of an RDE, which should provide a method of utilising the H or FJ cycle, in a much more compact form.

1.4 Rotating detonation engines

An RDE, such as the one shown as a cutaway in **Figure 3**, consists of an annular combustion chamber, into which fuel and oxidiser, either premixed or non-premixed, are fed through a series of orifices [3, 26, 36]. Each fuel/oxidiser mix requires a slightly different orifice geometry for optimal operation, so some devices have an adjustable injector plate [37, 38].

A detonation wave is initiated in the chamber, most commonly utilising a high speed flame that undergoes DDT by the time it enters the chamber [39, 40]. As this wave propagates around the chamber, it consumes the fuel, generating a high pressure zone behind it. This zone expands, and due to the geometric constraints, exits the chamber, generating thrust [35, 41]. An example of a CFD representation of the propagating wave can be seen in **Figure 4** [42]. Behind the wave, fresh fuel enters the chamber at a constant rate, priming that section of the chamber for the wave to continue on the next revolution, thus making a self-sustaining wave as long as fresh mixture is supplied [35, 43]. The detonation waves generally propagate close to the Chapman-Jouguet velocity (discussed in Section 3.2) for each fuel type (typically 1500–2500 m s^{-1}), so the effective operational frequency of current RDEs is approximately 1–10 kHz. Frequency is dependent on the chamber geometry, fuel,

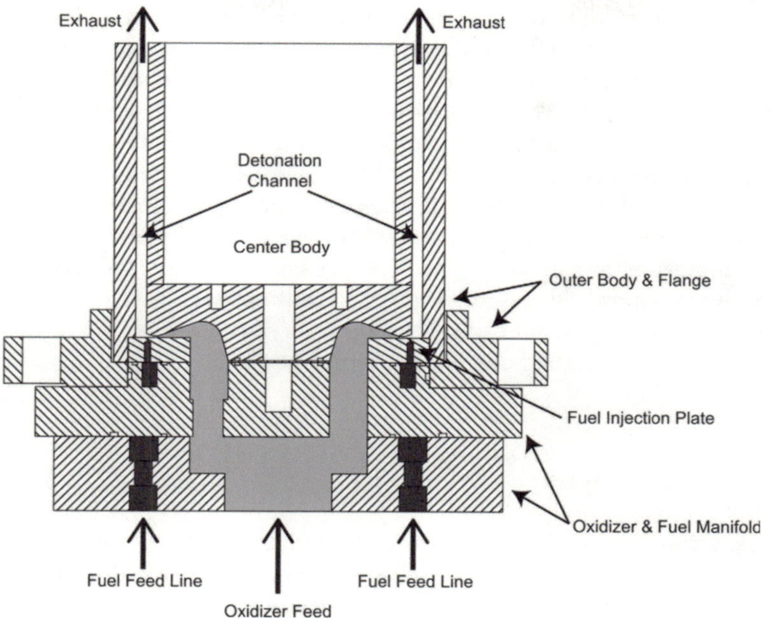

Figure 3.
Cross-section of a typical rotating detonation engine [38].

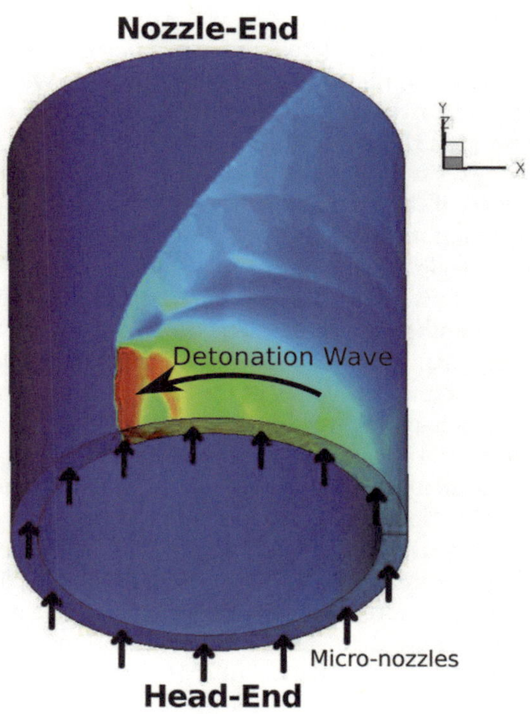

Figure 4.
3D model of the detonation wave propagation in an RDE [42]. The short arrows indicate the flow of fuel/oxidiser into the engine, and the long arrow indicates the direction of detonation propagation.

and thermal and frictional losses [31, 44]. The result is quasi-continuous thrust that approximates a continuous thrust through high frequency rotations, suitable for both direct propulsion applications and in the combustor of a gas turbine [31, 32, 45].

Important areas of RDE research include determining the wave characteristics, geometric constraints, the effects of pressure on the injection characteristics, determining fuel flow properties, and examining the geometry and structure of the detonation wave [3, 4, 30, 31, 41, 42, 44]. Additionally, there has been research into potential applications of detonation engines in which an RDE may be applied, such as air-breathing vehicles and gas turbines [46]. Despite a growing body of work on RDEs, there are still large gaps in current understanding that restrict practical application. Notably, optimising the system for wave stability, ensuring reliable detonation initiation, and ensuring the RDE does not overheat, are significant challenges facing engine development prior to commercial applications. Further development in this area would allow an engine to operate reliably over extended durations, with well-designed chamber and fuel supply.

2. Existing RDE designs

Most experimental RDEs are geometrically similar in design, consisting of an annulus made up of coaxial cylinders [5, 38, 47]. The chamber width, characterised by Δ, sometimes referred to as channel width, varies across designs. Several modular RDEs have been produced for testing various geometric parameters [30, 37, 48, 49]. As will be discussed in Section 4.4, the number of alternative designs to the annulus is limited. An exception is the hollow cylinder model to determine the effects of having no inner wall on the detonation wave as well as the practical feasibility [50].

There is reasonable consistency across published designs in the methods of initiating detonation waves in the RDE. Detonator tubes, in which a high-speed flame is encouraged to transition from deflagration to detonation, have been regularly and reliably used [26, 31, 32, 39, 49, 51]. It has been shown that the success of the detonation tube makes it an excellent initiator, producing a self-sustaining rotating detonation 95% of the time [26].

Like all jet-thrust reaction-based engines, the exhaust from a RDE may be channelled through a nozzle to increase thrust. Outlet and nozzle designs have varied across different RDEs. Many have not attached any nozzle, whilst some have chosen to utilise an aerospike [30, 31, 52]. The use of an aerospike increases performance through higher expansion area ratios, although the increased surface area results in higher heat flux and thus a loss of efficiency from the additional heat transfer [53]. Aerospikes may be directly attached to the end of the reaction chamber [31]. A diverging nozzle was found to increase the specific impulse, although the thrust increase was small, and for angles greater than 10°, the increase with angle was negligible [53]. None have made use of converging or converging-diverging nozzles, because the exhaust is typically flowing at supersonic velocities and thus could be choked through the converging cross-section. This would result in a loss of energy that would decrease the overall efficiency of the system.

A typical RDE, 90.2 mm in diameter, has been tested on a thrust sled [54]. It produced a thrust of 680 N using 176 g s^{-1} of C_2H_4/O_2 propellant at an equivalence ratio of 1.48 [54]. As can be seen from **Table 2**, this is well below that required for typical supersonic flight applications. The specific impulse (I_{sp}) of small scale operational RDEs has ranged from 1000–1200s depending on the fuel/oxidiser source used, though it is often H_2 with air [30, 31, 39, 41, 42]. The measured values of I_{sp} in these small scale RDEs are significantly below computationally predicted range: 3000–5500 s [31, 32]. However, a large scale RDE, discussed in further detail in Section 4, does operate with an I_{sp} of approximately 3000 s [5]. The experimental values for I_{sp} are similar to that of hydrocarbon-powered scramjets, but less than

Engine	Application	Thrust	Thrust to weight
Rolls-Royce Olympus 593	BAC Concorde	38,000 lb (169,000 N)	5.4:1 [55]
Teledyne CAE J402	McDonnell Douglas Harpoon	660 lb (2900 N)	6.5:1 [56]
Pratt and Whitney F135	Lockheed Martin F-35	191,300 N	11.47:1 (dry) [57]
General Electric F414-400	Boeing F/A-18E/F	98,000 N	9:1 [58]
Experimental RDE	None	680 N	3.47:1* [54]

**This is the thrust to weight ratio calculated using a pre-weight load cell system.*

Table 2.
Thrusts and applications of various engines.

turbojets and ramjets. These low values for small-scale RDEs are likely due to the use of unoptimised designs, and low chamber pressures [31].

RDEs have been found to be successfully operable with a range of gaseous fuels including hydrogen, acetylene and butane, as well as various jet fuels [30, 31]. Air, pure oxygen, and oxygen-enriched air have all be used as oxidisers [31]. Each of these has a variety of advantages and disadvantages, in both performance characteristics, and ease of obtaining, transporting, and storing the oxidiser. Particular difficulty is noted in the transport of gases such as H_2 and O_2 due to the high risk regarding transportation and significant compression of these chemical species [59]. In the case of transporting liquid fuels such as LH_2 and LOx cryogenic units are also required, adding to the already challenging process. The performance characteristics for several of these fuel types will be discussed further in Section 4.4.

The detonation wave velocity in operational H_2/air RDEs has been found to be on the order of 1000 m s^{-1} [30, 39]. In these RDEs, the operational frequencies are on the order of 4000 Hz, which produces quasi-continuous thrust [3, 32]. As wave speed is a key factor in the development of thrust, stable waves with high speeds are ideal for propulsion purposes. Stable detonation waves have reached maximum speeds in the range of 1500–2000 m s^{-1} in most designs using a H_2/air or H_2/O_2 fuel/oxidiser combination (more commonly the former), suggesting that there is open research into whether there is upper limit for detonation wave speed, and subsequently the thrust that may be produced [3, 22, 26, 60]. However, at very high frequencies (19–20 kHz), there may be multiple waves rotating around the annulus [60–62]. Multiple wave modes of propagation appear to be affected by fuel/oxidant equivalence ratio as well as total mass flow rate through the system. The high frequencies are a result of multiple waves travelling at approximately the same speed as the normal single wave. This phenomenon has the potential to provide more continuous thrust, though the higher frequency may limit I_{sp} due to insufficient refuelling of the detonation cell between waves. These wave modes have reliance on factors including fuel injection velocity, critical minimum fill height (discussed further in Section 4.3) as well as the detonation velocity [31]. Due to the inherent instabilities of rotating detonation waves, there are no specific relationships that can be determined between these factors and specific designs, only that they have an influence. Multiple wave fronts have been observed in several different RDE designs, where the general geometry has remained fairly similar [30, 31].

There are several methods of recording data from an operating RDE. Thrust generated may be measured with a thrust plate, and the flow rates of fuel and oxidiser may be measured or controlled within the supply lines [30]. The details of the shock may be recorded with pressure sensors attached to the chamber head, and external cameras [30]. Pressure sensors record the increased pressure generated by the shock, and by using multiple sensors, the detonation wave propagation velocity may be determined. A high-speed camera may be set up to capture the operation of the engine, allowing various parameters to be recorded, including the detonation wave propagation velocity, although this method is limited by spatial resolution, as the channel width can be quite small [30, 39]. A camera may also be used to image from the side, if the outer surface of the annulus is made of a transparent material [63]. Additionally, OH* chemiluminescence may be used to detect, record, and analyse the detonation waves in UV-transparent optically-accessible RDEs [64, 65]. These radicals are indicative of the reaction zone, and so, by analysis of their chemiluminescence, the structure of the detonation can be inferred. Often this detection is done through a quartz side window integrated into the RDE [63]. Peak intensity of the OH* chemiluminescence indicates the location of the detonation front, and so the effects of varying factors such as equivalence ratio and chamber geometries can be documented. Images are often phase-averaged and can by "unwrapped" for comparison to equivalent two-dimensional, "linearised", simulations and designs.

3. Detonation waves

3.1 Shocks

The structure of shock waves in gases was examined in detail by Voitsekhovskii in 1969, including those of shock waves in spinning detonations [66]. These examinations resulted in the first diagram of the structure of a spinning shock wave, and the identification of a number of features, which are identified from the computational model of an RDE shown in **Figure 5** [32]. This model used premixed hydrogen/air as the fuel/oxidiser mixture and has been "unwrapped" into two-dimensions (this approach is described in Section 5.1). Feature A is the primary detonation front; Feature B is an oblique shock wave that propagates from the top of the detonation wave; Feature C is a slip line between the freshly detonated products and older products from the previous cycle; Feature D is a secondary shock wave; Feature E is a mixing region between the fresh premixture and the product gases, where deflagration may occur [67]; Feature F is the region where the injector nozzles are blocked; and Feature G is the unreacted premixture.

In both **Figure 5b** and **Figure 8c** (Section 4.3) the detonation cell structure can be seen, with high pressure zones outlining each cell. These lines of high pressure contain triple points, where the transverse and oblique shocks meet the Mach stem of the detonation wave [68, 69]. The concentrated pressure at these triple points is the point of maximum energy release, and the subsequent pressure spike when two triple points collide generates new detonation cells [68, 70]. While this generation is the main reason behind the propagation of detonation waves, the triple points still require further investigation as to the effects they have on the overall characteristics of a detonation wave [70]. The direction of these triple points can be seen as the white lines in **Figure 8c** with trailing high pressure zones forming the walls of the detonation cells. As the detonation cell width is defined by the geometry of the system and the chemical composition of the detonating fuel, it seems that the triple point velocity and direction must also directly relate to these factors, although limited research has been done to formally connect these points.

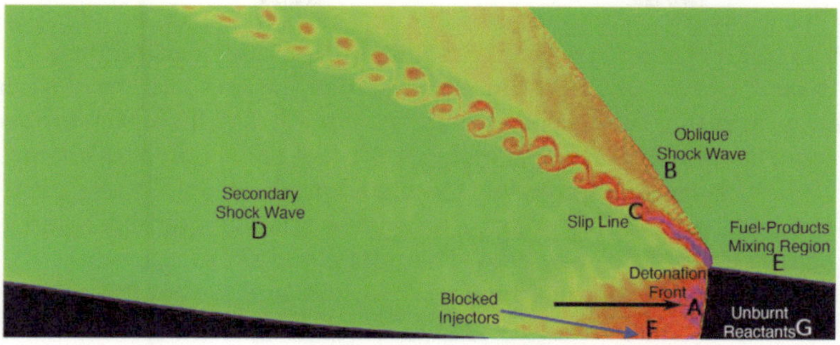

(a)

(b)

Figure 5.
Pressure contour indicating the cell structure of a detonation wave in an RDE with a premixed supply, taken from a computational modelling study [32]. (a) Pressure contour indicating the full structure of detonation in an RDE, "unwrapped" into two dimensions. Feature A is the detonation wave, Feature B is the oblique shock wave, Feature C is the slip line between the freshly detonated products and products, Feature D is a secondary shock wave, Feature E is a mixing region between the fresh premixture and the product gases, Feature F is the region with blocked injector nozzles, and Feature G is the unreacted premixture. The arrow denotes the direction of travel of the detonation wave. (b) A close-up image of the detonation front.

In an RDE, the detonation wave remains attached to the base of the annulus, as illustrated in **Figure 5b** and in **Figure 6** [3, 6, 71]. This is due to the continuous fuel/oxidant supply [3, 71], as a premixture or allowed to mix in the chamber ahead of the detonation wave [32, 39]. There is also some evidence that stable, lifted waves may also be possible if there is insufficient mixing between the fuel and oxidant

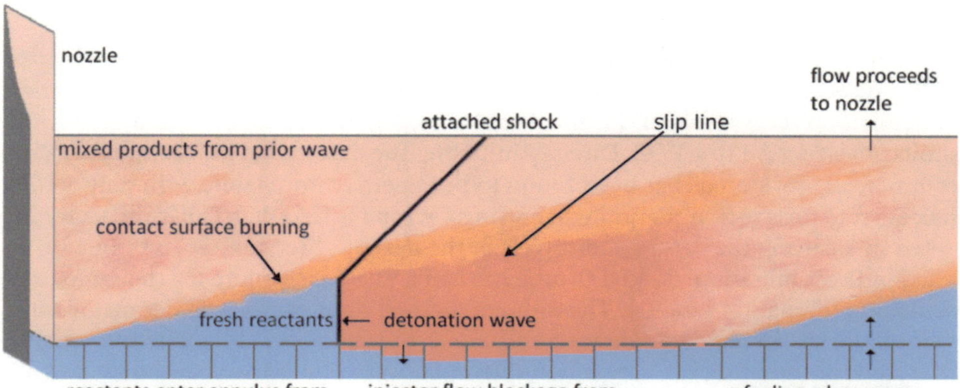

Figure 6.
Diagram showing the general structure of the detonation in an unwrapped RDE [3].

[27, 44]. The propagating detonation wave combusts the reactants [32, 39] which generates a region of extremely high pressure immediately behind the wave. This pressure is on the order of 15–30 times higher than the pressure ahead of the detonation, preventing flow through the injectors [3]. The high pressure zone expands in a Prandtl–Meyer fan, allowing fresh fuel and oxidiser to enter the chamber [35]. This expansion propels the mixed products axially along the engine, generating thrust. In addition to the primary shock, an oblique shock and a secondary attached shock are also generated (Features B and D in **Figure 5a**).

At the interface between the premixed reactants and the combustion products, there is a significant difference between the conditions of the unburnt fuel/oxidiser mixture and the products. This causes some deflagration along the slip line, as shown in **Figure 6**, generating Kelvin-Helmholz instabilities, which vary the detonation propagation velocity [3, 22, 72, 73]. This decrease in the propagation velocity results in an increase in the pressure, disturbing the oncoming shock wave and forcing the sonic flow directly behind the shock wave to undergo supersonic flow acceleration [74]. As shown in **Figure 6** there is a section of injector flow blockage that occurs as the wave passes the fuel array. The high pressure front from the shock wave causes stagnation of the injector flow, or even back-flow which, if not handled, could cause catastrophic failure of the system [3, 6, 36]. This back-flow is a strong reason as to why the fuel and oxidants should not be premixed in practical systems or experimental investigations as it can result in flashback.

3.2 Shock initiation

The Chapman-Jouguet (CJ) condition can be defined as the requirements for the leading shock of a detonation to not be weakened by the rarefactions of the upstream detonation products [75]. This sonic plane then acts to allow the supersonic expansion of the detonated gases to occur without disturbance by rarefactions downstream of the flow [75]. The CJ condition can be used to approximate the detonation velocities in three-dimensional models but is better suited to a one dimensional analysis with an infinitesimally thin detonation front [76]. Despite this, it is used in most instances of numerical modelling as a guide as to whether the wave is performing as expected for the given parameters of the RDE [4, 6, 27, 31, 32, 42, 75, 77]. Chapman and Jouguet's theory only applies to kinetic energy, disregarding the chemical energy of the reacting species, and hence, the Zel'Dovich-von Neumann-Doring (ZND) model is used as a more complete representation of the shock, taking into account the finite chemical reaction area directly upstream of the leading shock [3, 21, 45, 75, 78–80].

There are two methods which may be used to initiate the detonative shock in an RDE—directly in the chamber, or indirectly via a high speed flame in a deflagration to detonation transition (DDT) tube [26, 31, 39, 49, 51]. These tubes are very similar in structure to a PDE. Directly initiating the detonation in the chamber via commercial spark plugs has been found to be generally unreliable, with only a 40% success rate for shock initiation when using CH_4 in O_2 [26]. Particular difficulty is noted in ensuring the detonation travels in the desired direction [26, 32]. In contrast, indirect initiation via a DDT tube has had a 95% success rate for the same fuel/oxidant combination [26, 31]. The indirect method involves using a detonator tube that can be set up in any orientation relative to the chamber, although tangential is favoured for initiating the detonation direction. Initiation is then caused by a small volume of a highly detonative mixture being ignited by spark plugs before DDT occurs, thus initiating the RDE. Perpendicular initiation can also be used, but this often results in the development of two detonation waves that rotate around the chamber in opposite directions [31]. Collision of these opposing waves usually destabilises the system as the waves weaken and reflect back in the direction of origin [31]. Desired direction also appears to be affected by initial total pressure and ignition distribution around the fuel plenum [27, 81]. For a desired single wave direction and propagation, tangential initiation is the most suitable method. Although slightly less compact due to the initiator tube, this may be reduced by placing obstacles in the tube to accelerate the DDT, or by using a more detonative fuel than that used in the primary process [31, 48, 62, 82, 83]. Using an initiator tube, however, may produce small wavelets ahead of the main detonation front, which, if present, reduce the detonation propagation velocity by up to 60% [84]. Once the main detonation is running, the interface between the initiator tube and main chamber must be closed off prior to the shock completing a revolution of the chamber [84]. Additionally, there may be a slight delay, on the order of milliseconds, between the detonation exiting the DDT tube and the commencement of full RDE operation in order to purge the spent reactants from the DDT process [85]. This delay seems to only be transient with no large effects on shock structure or stability, and the excess products are expelled along with the rest of the exhaust [85].

3.3 Instabilities

Three-dimensional modelling has shown that increasing the width of the channel—whilst maintaining the equivalence ratio, injection pressure, chamber length, and injector configuration—increases the detonation velocity, but the transverse shock wave ceases to be aligned with the radial direction [22, 27, 86]. As can be seen in **Figure 7**, the point of contact with the inner wall begins to lead the detonation wave as the channel width increases [22]. This phenomenon generates reflected shocks from the outer annulus wall, which may produce instabilities in the primary shock. It has been suggested through qualitative observation, however, that the effect of upstream reflected shocks on the shock structure may only be minimal [39, 87]. Once the channel becomes sufficiently wide, as shown in **Figure 7c**, the shock wave detaches from the inner wall, briefly forming a horseshoe shape against the outer wall [22]. This allows significant amounts of fuel to pass through the engine without combusting, and produces large instabilities and fragmentation in the detonation wave, which causes the structure to collapse [22]. These lead to a significant loss of performance, and secondary detonations in the exhaust [22]. It has been noted that increasing the channel width also results in increased variance of I_{sp}, and that, combined with high fuel flow rates, leads to the formation of secondary waves, which in turn leads to hotspots and choking the fuel supply

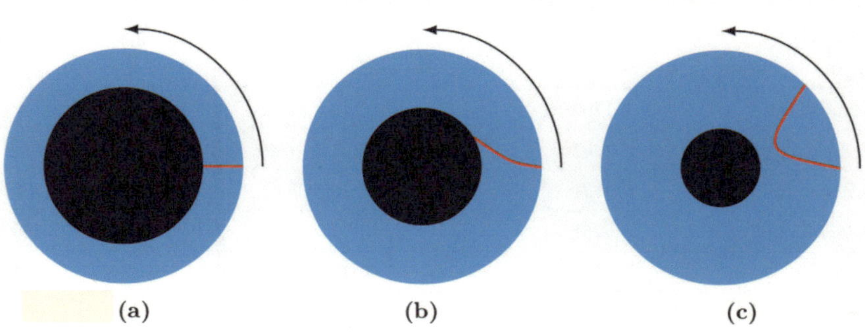

Figure 7.
Schematic of three different RDE designs showing the effect of varying the channel width on detonation structure. Arrows show detonation wave propagation direction. The red line is detonation wave, indicative only. Based on research from [22]. (a) Narrow channel, (b) mid-sized channel, and (c) wide channel.

[42, 62]. This is likely due to the increase in size of the interface area producing greater Kelvin-Helmholz instabilities, resulting in larger variances in the detonation velocity [42].

It has been found that using a fuel-rich mixture produces stable waves with high detonation velocity and efficiency [80, 88]. Higher mass flow rates have also been attributed to increasing the chance of a stable wave being formed [6, 89]. Additionally, it has been shown that the equivalence ratio has a strong influence on the effectiveness of detonation and the stability of the system [80]. Detailed investigation has shown that the stability of the system is improved with increased equivalence ratio, but indicated a maximum equivalence ratio of 1.27, before the detonation wave became short-lived and transient, which is unsuitable for practical purposes [60]. Whether this is a universal limit, or a limit of that particular investigation is unclear, and requires further research. Furthermore, the findings indicated that lower equivalence ratio influences the number of wave fronts produced, with stoichiometric seeming to be a transition point to a stable one wave propagation mode [60, 86, 90]. It is interesting to note that for lean mixtures, the initial channel pressure needs to be higher for a stable detonation to propagate [88].

4. Factors influencing the design of RDEs

4.1 Fuel

The wave propagation velocity varies with the fuel/oxidiser combination. A variety of mixtures have been tested in a detonation tube of an RDE, with their wave propagation velocities and wavefront pressures shown in **Table 3**, which is indicative of their varying performance in an RDE. It should be noted that the pressure, energy and specific impulse in **Table 3** are determined with a detonation tube, and provide a numerical comparison between each fuel/oxidiser combination. Hydrogen/oxygen mixes have been ideal for modelling purposes due to the simple chemistry involved, and are often used in experimental work due to the predictable behaviour. Additionally, the high detonation propagation velocity and wavefront pressure of hydrogen makes it a suitable fuel for real applications. Another common fuel choice is methane, due to the satisfactory propagation velocity and specific impulse in testing [31]. As mentioned in Section 2, the theoretical I_{sp} is still greater than that of a standard turbojet propulsion system, irrespective of fuel selection [91].

Fuel mixture	Detonation speed (m s^{-1})	Wavefront pressure (atm)	ΔH_r(MJ kg^{-1})	I_{sp}(s)
Hydrogen/ oxygen	2836	18.5	8.43	289.39
Hydrogen/air	1964	15.5	3.48	200.41
Ethylene/oxygen	2382	31.9	5.23	243.06
Ethylene/air	1821	18.2	2.85	185.82
Ethane/oxygen	2257	29.0	4.87	230.31
Ethane/air	1710	15.8	2.49	174.49
Propane/oxygen	2354	34.2	5.18	240.20
Propane/air	1797	17.5	2.80	183.37

Table 3.
Fuels, wave propagation velocities and pressures, heat of combustion (ΔH_r), and specific impulse I_{sp} [36].

Transportability of fuel, and maintenance of fuel lines, are deciding factors in determining which fuels can be used. These issues are especially important for aerospace applications. Gases such as H_2 and O_2 are particularly volatile and reactive, hence can be difficult to transport in the large quantities needed for use in an RDE. Therefore, gaseous fuels and non-air oxidisers are challenging and largely unsuitable for real world applications [5]. However, H_2 does have a high heat of combustion that is not matched by liquid hydrocarbon fuels. Jet fuel, kerosene, octane and other long-chain hydrocarbons provide a practical alternative to the H_2/O_2 mixture though. High volumetric energy density as a result of liquid state, as well as greater ease of transportability makes these hydrocarbons a more feasible fuel choice.

There are several issues regarding fuel choice that deserve further discussion. In particular, the use of cryogenic fuels for cooling the engine is a beneficial approach, increasing thermal efficiency, as well as reducing the thermal load on other components such as mounting systems [3]. Another advantage is a higher volumetric energy density that comes from the compression of normally gaseous fuel sources. Testing of liquid oxygen (LOx) and gaseous or liquid hydrogen (GH_2/LH_2) fuel/oxidant systems for viability has been performed, but implementation in real world scenarios is challenging [92, 93]. Liquid hydrocarbons require further investigation to demonstrate their effectiveness in producing thrust through detonation [30], particularly because of the need for flash vapourisation to avoid multiphase effects in the mixing process [30, 51].

4.2 Injection

An axial fuel injection process through a circumferential orifice plate was consistent across most simulations and real world models as an injection scheme [5, 6, 22, 26, 30, 32, 36, 38, 39, 41, 42, 42, 52, 61, 62, 82, 86, 88, 92, 94–99]. Further research is required into fuel blockage effects due to the high pressure of the shock wave, with particular emphasis on the effects of increasing fuel pressure to alleviate blockage and increase overall engine performance [100]. In the majority of numerical and physical models, such as **Figure 3**, fuel and oxidiser are injected through an orifice place around the annulus, allowing them to continually feed the propagating detonation wave. Typically, the fuel and oxidiser are fed in separately, and allowed to mix in the chamber [26]. This design is also used in most numerical models, although some have used premixed fuel/oxidiser as a simplified boundary condition. Almost all physical designs have been built without a premixed fuel/oxidant injection scheme due to concerns with flashback [99]. In a premixed design, the shock wave may propagate into the injection plenum, carrying with it the reaction front. With sufficient pressure though, typically 2.3–3 times the chamber pressure, this can be avoided [32].

Investigation into flow characteristics of a turbulent inflow have shown that there are specific zones within the chamber which favour different forms of combustion: some zones favour deflagration, and others favour detonation [101]. The larger deflagration zones created reduce the thermodynamic efficiency of the engine, indicating that fuel flowrate influences the reliability of an RDE [101]. It has been suggested that high inlet velocities generate incomplete combustion and hot spots, reducing detonation wave stability and reducing system efficiency, although further research is required [102]. As indicated in Section 3.3, the introduction of instabilities in the flow profile can decrease the efficiency of the engine as well as disrupt the detonation wave itself. Further findings indicate that increasing the fuel injection area, particularly by increasing the number of orifices, results in more efficient pressure gain [86, 97, 99, 103]. This produces a larger expansion wave of the previous combustion reactants, generating higher thrust, without disrupting the flow-field characteristics [98]. However, with lower fuel injection velocities comes an increased risk of flashback. There is, therefore, some optimal fuel injection area for operation which requires further work to verify [98]. Finally, the pressure ratio between the inlets and the engine outlet also has an effect on the I_{sp} of the engine, with pressure ratios of less than 10 showing notable reductions in impulse [32, 72]. Thus, because of these conflicting requirements, injector design is complex and more research is required such that fuel consumption and thrust output are optimised.

4.3 Scalability

Existing RDEs tend to be relatively small, and therefore may need to be scaled up, or arranged in parallel, to produce thrust required for practical applications, such as those listed in **Table 2**. One method of scaling RDEs is to run multiple identical devices in parallel, in a similar manner to that used to run multiple PDEs [34, 104]. However, this would require more complex plumbing, increasing the weight of the overall system, and thus decreasing the thrust-to-weight ratio. However, this solution has not been explored in any depth and its viability is unknown.

In order to make larger RDEs, in-depth research into the geometry of the combustion chamber is required. A number of relationships between the critical detonation wave height and the various dimensions have been identified [27, 30]. Detonation structure, as described in Section 3.1 is composed of small diamond shaped detonation cells that make up the front. The widths of these cells are dependent on the energy of the detonation (related to the fuel in use) as well as the available geometry for detonation. In this way, the equivalence ratio can be a large determining factor [30, 105, 106]. Critical minimum fill height is the minimum mixture height required for a detonation wave to propagate through a given fuel/oxidiser mixture. It has been found that the critical minimum fill height, h^*, and the minimum outer wall diameter, $d_{c_{min}}$, are related to the detonation cell width, λ, by

$$h^* \propto (12 \pm 5)\lambda \tag{4}$$

$$d_{c_{min}} = 28\lambda \tag{5}$$

and the minimum channel width, Δ_{min} is related to the h^* by

$$\Delta_{min} \propto 0.2h^* \tag{6}$$

Finally, the minimum axial length of an RDE, L_{min} is related to the actual fill height, h, by

$$L_{min} = 2h \tag{7}$$

although lengths under 2–3 times the minimum result in reduced efficiency due to incomplete combustion [27]. However, in simulations, it has been suggested that for low inlet-nozzle pressure ratios the wave the wave height grew with the chamber length, reducing the I_{sp}, of the engine [42]. For high pressure ratios, no such reduction was indicated [42]. **Figure 8** indicates the physical representations of the above variables.

There is not yet any theoretical data for λ, but there are multiple models which may be used to predict the value under various conditions [78]. It is known that more highly reactive mixtures, such as H_2/O_2, have lower λ values, and so have minimum

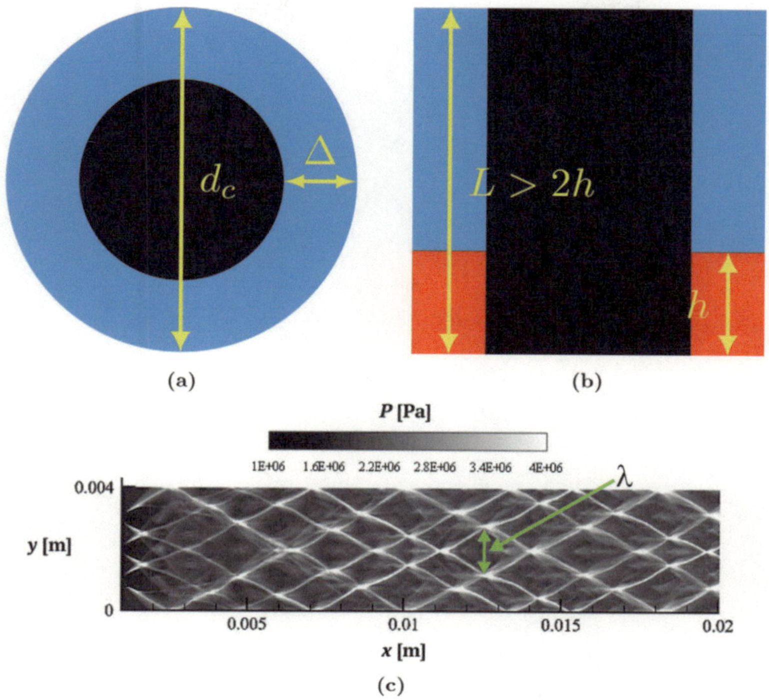

Figure 8.
Geometric parameters of an RDE. The red area is the area filled by the fuel/oxidiser mix in which the detonation propagates. (a) Top view, (b) side view, and (c) detonation cell width adapted from [79].

chamber diameters on the order of 40–50 mm. Liquid hydrocarbons, such as kerosene and jet fuel, combusting in air, have reactions with higher λ, so, when Eq. (5) is applied, the minimum chamber diameter is calculated to be 500 mm [3].

Modelling a large-scale RDE presents a challenge due to increasing computational requirements with increasing size, so limited work has been done in this area. Nevertheless, a larger scale experimental RDE has been demonstrated [5]. This RDE had an outer chamber diameter of 406 mm, and a channel width of 25 mm, and an air inlet slit that could be varied across the range 2–15 mm [5]. It produced a consistent thrust of 6 kN with a combined fuel/oxidiser flow rate of 7.5 kg s^{-1}, whilst also producing an I_{sp} at 3000 s, consistent with the computational models noted in Section 2 [5, 31]. This is approximately four times the physical size, $\sim$40 times the consumption of combined fuel/oxidiser, and $\sim$12 times the thrust of other RDEs noted in Section 2 [46, 54]. Although still producing low thrust compared with conventional jet engines, such as those listed in **Table 2**, it is also half the diameter of the modern engines [57, 58]. Furthermore, 6 kN would be more than sufficient thrust for use in a Harpoon missile [56], and this RDE shows that they are capable of being scaled beyond small sizes.

4.4 Alternative designs

The design used in most simulations and experimental work is a coaxial cylinder structure [3, 27, 31, 35]. This simple geometry is advantageous for both modelling and manufacturing. Design variations including using nozzles, aerospikes such as that shown in **Figure 9**, or an entirely hollow cylinder, have been utilised in several RDE designs [5, 52].

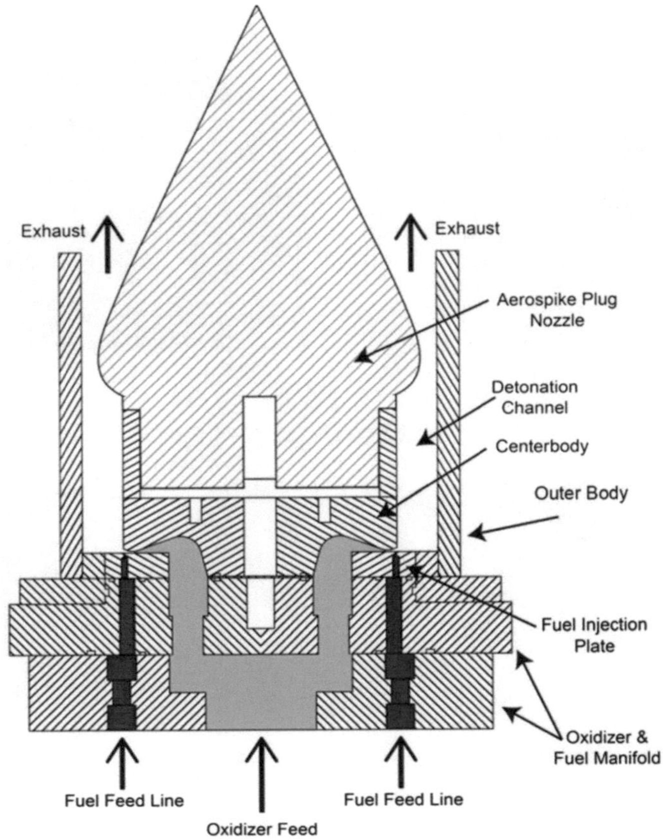

Figure 9.
Example of an aerospike nozzle configuration [52].

Alternative chamber geometries have been largely limited to adjustments in the diameters of the chamber [4, 42], including with different sized engines [15, 31, 39, 54]. Other work has been conducted on a single RDE with interchangeable outer wall sections [22, 30]. As noted in Section 2 and Section 3, both of these factors influence the stability and the performance of RDEs. The effect of varying the length of the chamber on the detonation propagation has been investigated, which led to the previously mentioned requirement that the chamber be at least twice, and preferably four to six times, the fuel fill height [4, 96].

Hollow RDEs, dubbed "centrebodiless" designs, have been tested with two different designs [50, 61]. One design was identical to a conventional RDE 100 mm across, but the inner cylinder terminated parallel to the fuel/oxidiser injectors [61]. In this design, tested with 169.7 g s^{-1} of CH_4/O_2 at an equivalence ratio of 1.154, it was found that the detonation was unstable [61]. The fuel and oxidiser were free to move into the space usually occupied by the centre body, and thus insufficiently mixed to sustain a stable detonation [61]. However, when the same geometry was tested with 253.3 g s^{-1} of CH_4/O_2 at an equivalence ratio of 0.665, the mixture became sufficiently mixed to sustain a stable four-wave detonation structure [61]. Another design was completely hollow, allowing oxygen-enriched air to be pumped through the centre of the chamber, and fuel was supplied around the edge [50]. In this design, stable detonations, operating at ~8000 Hz were achieved at an equivalence ratio of ~0.4 [5-]. However, this design required that the molecular ratio of nitrogen-to-oxygen in the oxidiser be approximately two for detonation. Nitrogen-to-oxygen ratios of ~2.5 produced deflagration, and a ratio of 3.75—approximately standard air—led to the RDE self-extinguishing [50]. Nevertheless, the need for oxygen enrichment introduces additional cost and challenges for practical RDEs in propulsion applications. It was also noted that the oxidant flow provides an outward pressure that acts like a wall but carries no extra weight, and even adds a small amount of thrust as the air is expelled [50]. Both designs can be looked at as successful proofs of concepts, and potential first steps in simplifying the geometry of an RDE, with the latter being potentially useful in applications such as after-burners [50, 61]. However, this concept has not been explored with pre-heated reactants, such as those which would be present in an afterburner.

The attachment of turbines to RDEs has been proposed [8, 9, 31, 32, 45]. It has also been noted that there is a secondary shock propagating from the detonation, which exits the outlet of the chamber [32]. However, turbine blades are sensitive to shocks. As such, the effect of the secondary shocks on the blades of potential turbines must be investigated. It is worth noting that an experimental PDE array has been tested with an attached turbine, in the form of an automotive turbocharger [31]. In that case, a buffer chamber was inserted between the PDE and the turbine [31], and such a technology may be suitable for RDEs.

5. Modelling and development tools

5.1 Planar and three-dimensional modelling approaches

Computational fluid dynamics (CFD) modelling is a powerful tool for the analysis of rotating detonations prior to, or in tandem with, experimental systems. The majority of numerical studies have aimed to provide in-depth understanding and details of the detonation structure [22, 41, 62, 67, 72, 94, 107, 108] or assess the physical and modelling factors influencing performance [32, 67, 73, 109].

Computational models of the azimuthal detonations in RDEs may use full three-dimensional geometries [20, 22, 67, 94, 95, 107, 110] or simplified, two-dimensional geometries [6, 32, 41, 43, 62, 72, 73, 108, 109, 111–114]. The former, higher-fidelity, approach can incorporate complex geometric and flow features, although require $\sim$ 10,100 million numerical cells for high fidelity large-eddy simulations (LES) or direct numerical simulations (DNS) [22, 94, 95, 112]. These may subsequently result in considerable computational expense in conjunction with detailed turbulence and combustion chemistry. In contrast, by assuming that the channel width is much smaller than the diameter, the annulus geometry may be "unwrapped" [108] and treated as a planar flow [41]. The azimuthal detonation repeatedly travels through the domain using periodic boundaries (i.e. the outflow from one side feeds into the other side). Such a model was shown previously in **Figure 5a** [32], where the detonation is travelling left-to-right and the two vertical edges of the image are the periodic boundaries. This can be seen by noting the height of the unreacted premixture region (Feature G) at each side of the figure. The stationary geometry shown in **Figure 5a** [32] shows a full, two-dimensional, unwrapped RDE geometry, and allows the detonation to freely—and repeatedly—propagate through the domain. It may, in some cases, be beneficial to examine the detonation in its own frame, by matching the domain velocity to the negative of the detonation speed; however, this requires significant trial-and-error as the detonation speed cannot be accurately approximated as the CJ velocity for this purpose [108].

Two-dimensional modelling of RDEs assumes that the flowfield along the centre of the channel is representative of shock and deflagration structure across the entire width. Consequently, this inherently assumes slip-wall conditions and that the detonation-front is normal to the two-dimensional geometry. In the unwrapped two-dimensional geometry, all fuel is injected axially from one edge (the bottom edge in **Figure 5a** [32]) and is exhausted through the opposite edge (the top edge in **Figure 5a**) [6, 32, 72, 111]. It therefore follows that all exhaust products must leave the domain axially, due to conversation of angular momentum. This was confirmed in early two-dimensional modelling, which found that the density-averaged azimuthal velocity was less than 3% of the axial velocity [41]. Such a criterion could be extended to assessing whether a three-dimensional model, at some fixed radius within the channel, could be treated as an unwrapped planar domain.

Detonation wave curvature, imperfect mixing, three-dimensional turbulent structures and transverse shocks are features reported in three-dimensional computational modelling [22, 67, 79, 94, 107] and experimental studies [62]. These features arise from the effects of channel size [22], discrete injectors [79] and interactions between transverse waves and walls [62, 79]. These features are inherently three-dimensional and cannot be captured using planar, periodic models, and require more complex computational geometries.

5.2 Boundary conditions in computational models of RDEs

Fuel/oxidiser inlets may be modelled as simple points, lines, surfaces or complex, discrete injectors. The latter may be treated as a series of inlets in two-dimensional models, assuming upstream micro-mixing [109, 112]. Differences in the injector configuration can lead to differences in detonation pressure [112], or lifted flame behaviour in the event of poor mixing in a partially premixed system [109]. The study which observed the latter phenomenon, however, was undertaken using the Euler equations, which may affect the fidelity of modelled mixing (discussed later in this section), and a simplified induction parameter model (described in Section 5.4) [109], although this has also been observed experimentally in C_2H_2-fuelled RDEs [115].

Inlet boundary conditions in premixed models, are often defined by inlet throat-to-nozzle-exit ratios. These, and the set upstream pressure, control whether the inlets are blocked, subsonic or choked and are chosen to range from 0.1–0.2 [6, 109, 110, 112], although ranges as large as 0.07–0.3 have shown little effect on I_{sp} [73]. More complex fuel injector geometries have been assessed through three-dimensional modelling [94], demonstrating the effects of the complex detonation/deflagration interactions on imperfect mixing, however, neither instantaneous (fuel or air) plenum pressures nor detonation wave-speeds could be correctly predicted.

5.3 Turbulence modelling in RDE simulations

Rotating detonation engines have often been numerically modelled using the compressible Euler Equations [6, 20, 32, 41, 43, 62, 72, 95, 108, 110–112]. The Euler equations conserve momentum, mass and energy, but do not account for viscosity, following the assumption that the detonation structure dominates viscous dissipation. Viscous effects may, however, be incorporated into numerical studies of RDEs through the use Reynolds-averaged Navier Stokes (RANS) modelling [107, 113], LES, LES-RANS hybrids such as [improved] delayed detached eddy simulations (IDDES) [67, 94], or DNS [22]. Of these approaches, Euler, IDDES and DNS studies [22, 41, 67] have all been able to capture Kelvin-Helmholtz instabilities in the unreacted/reacted and the post-shock mixing layers (see **Figure 5a** as an example), using sufficiently small element sizing in both two- and three-dimensional models.

The grid required to resolve large structures in RDE mixing layers is dependent on the size of the geometry. Elements of 200 μm have been shown to predict shear layer instabilities using either Euler equations or IDDES in an RDE with a mid-channel diameter of 90 mm [67] and an ~140 mm inner diameter RDE required axial and azimuthal elements smaller than 200–300 μm to capture the structures in a DNS study [22]. In contrast, Kelvin-Helmholtz structures were not observable in models of a 1 mm outer diameter RDE with computational elements larger than 1.25 μm [73]. In all cases, these minimum azimuthal element sizes are $\lesssim$0.21% of their respective mid-channel diameters, suggesting a minimum relative element size relative to geometry. These element sizes are not, however, proportional to the CJ induction lengths which are ~200–300 μm for stoichiometric H_2/air mixtures near 300 K [116, 117], compared to ~50 μm H_2/O_2 [117].

Both viscosity and species diffusion have been stated as critical features in non-premixed models of RDEs, promoting the use of IDDES or LES in modelling studies [67]. In contrast, a negligible dependence of detonation velocity or I_{sp} was reported in DNS of a partially-premixed "linearised" model [114] (refer to Section 5.5 for more on these models). Despite this, it is crucial to note that Euler equation models significantly over-predicted deflagration upstream of the detonation in the premixed numerical RDE model [67], whereas the mixture upstream of the shock in the linearised model is completely unreacted [114, 118]. This warrants further study on the differences of these modelling approaches on detonation interactions with non-premixed fuel/air injection into post-combustion gases. This is further complicated by the suggestion that the absence of viscous dissipation and diffusive mixing in the Euler equations could enhance perturbations driven by baroclinic vorticity generation which is, in turn, promoted by wrinkling in the deflagration upstream of the detonation.

Although the Euler equations cannot account for viscous effects, such as wall shear-stress and heat transfer, these have a small, but non-negligible, effect (~7%) on predicted I_{sp} compared to IDDES modelling including non-slip, isothermal walls in premixed RDE models [67]. The appropriate selection of wall boundary conditions will therefore likely prove to be an important factor in RDE development, with different thermal treatments significantly changing the fraction of fuel burnt upstream of the detonation wave [67]. Neglecting these physical features, results in decreased deflagration away from the detonation wave, with adiabatic walls most significantly over-predicting combustion outside of the detonation wave [67]. Despite this, detonation wave-speeds were reasonably insensitive to wall temperatures in the range of 500–800 K in the same study, and consistently over-predicting experimentally measured detonation wave-speeds [94], although temperatures significantly exceeding the autoignition temperature (up to the adiabatic wall temperatures ~2000 K) were not assessed.

Incorporating viscosity and thermal wall-effects into IDDES simulations requires significant computational resources. One such study required a computational mesh of ~100 million computational elements, included multiple chemical species and reactions, with numerical time-steps of 30 ns [94] and is similar to an earlier study using approximately one-third of the number of cells which required ~ 35,000 CPU-hours to solve [67]. Several cases in an earlier study, however, required ~9 million CPU-hours to produce a final solution due to the use of time-steps of 2 ns [67]. In addition to IDDES studies, viscous and diffusive effects may be accounted for in unsteady RANS modelling [107] and facilitate the inclusion of detailed chemistry (see Section 5.4) with significantly lower computational overhead than IDDES or DNS. Such RANS models cannot, however, capture the turbulent fluctuations in the instantaneous flow-field, although there is evidence that they may be able to provide sufficient accuracy for parametric studies of mixing, detonation wave structure and loss mechanisms in RDEs [119, 120]. The interactions between detonations, deflagration and viscous and thermal wall-effects add further complexity to producing RDE models which can accurately reproduce experimentally measured engine characteristics, although the computational resources may currently prohibit broad parametric studies using high fidelity modelling approaches.

5.4 Chemical kinetics and interaction models

The majority of numerical RDEs works to date targeted H_2/air and H_2/O_2 systems [6, 20, 22, 41, 62, 72, 73, 79, 94, 95, 111, 112, 118, 121, 122], given their relatively simple chemistry in comparison with both small and large hydrocarbons. Nevertheless, limited data are also available for linearised CH_4/air and C_2H_4/air systems [114].

The simplest approach to describe the chemistry is that of a one-step irreversible reaction [6, 43, 62, 95, 108, 109]. This assumption has been widely used to numerically investigate various aspects of fully premixed canonical RDE cases and useful insights have been gained [6, 32, 95]. However, it is well known that such a simplification is not able to accurately quantify many detonation responses of interest (e.g. upstream deflagration phenomena [109], triple shocks structure [79, 116]), mainly due to the sensitive Arrhenius nature of the reaction rate to temperature variations. Also, the use of *ad hoc* correlations of the experimental data with adjustable kinetic parameters (e.g. reaction order, activation energy) are only valid for a limited range of the system and thermodynamic parameters [116].

Simplified approaches to chemical kinetics may employ a one-step reversible reaction [20, 62] or a two-step mechanism [22, 41] to describe the chemistry within a system. In particular, for the one-step case, the forward reaction rate is calculated using the classical Arrhenius equation with the reaction rate constants tuned from a reference case while the backward reaction rate is calculated from the assumption of local chemical equilibrium [20, 62]. This approach has been validated against detailed chemistry for a 1D model [20]. For canonical 2D premixed RDEs, a one-step reversible reaction is not able to accurately capture the post-detonation temperature while it is able to predict both the experimental pressure and velocity fields [20]. In addition, it was also found that this approach can be successfully implemented to describe stratification effects in three-dimensional non-premixed RDE systems [62].

For the one-step case, a number of two- and three-dimensional premixed RDE simulations employ an induction-time parameter model (IPM) to compute the chemical source terms [6, 32, 43, 109]. The IPM has shown reasonable accuracy for the prediction of detonation wave propagation in premixed systems [108], as the induction time is derived from the same configuration as the CJ wave-speed [116]. In addition, it is computationally inexpensive as a global induction parameter allows for release of energy over a finite period of time. Nevertheless, the IPM lacks the flexibility to accurately describe the physics occurring in more realistic non-premixed systems [94]. The thermodynamic properties of the single product species employed in this model are dependent upon the equivalence ratio of the fuel/air mixture. Therefore, this approach cannot easily handle the spatially varying local equivalence ratio occurring in a non-premixed system [116]. This model also lacks the capability to capture the low-pressure heat release and the change in equilibrium chemistry of post-detonation products. Finally, this method requires *a priori* calculation of the CJ induction time, but the computed detonation velocities in detailed simulations can be significantly higher than that of CJ velocity [94]. If this approach is extended to a two-step reaction model (consisting of an induction reaction followed by an exothermic recombination reaction), two progress variables are obtained and need to be solved in lieu of individual species concentrations. This approach is termed two-parameter progress variable, and it has been successfully applied for premixed systems [22, 41]. Nevertheless, the variation of the two source terms is extremely sensitive to the choice of the constants adopted [22]. Global chemistry has also been implemented through the well-known PDF method [107], although this approach is generally used for detailed chemistry in combustion processes [123].

Finite-rate kinetics and the associated kinetic mechanisms are needed to capture complex phenomena such as near-limit propagation leading to quenching of the detonation wave [116]. This is mainly because the use of a one-step reaction precludes the influence of chain-branching-termination mechanisms that are invariably multi-step in nature. In this regard, an advanced approach is the induction-length model, which concerns determining the induction length for adiabatic propagation and using it to estimate global detonation parameters such as the cell size of steady propagation and the wave curvature at quenching [116]. This study showed that at least a four-step mechanism is required to achieve acceptable predictions in CJ detonation.

Models of RDEs using H_2/air, H_2/O_2, CH_4/air and C_2H_4/air mixtures have employed detailed chemistry and simplified configurations [68, 72, 73, 79, 111, 112, 114, 118, 122], although only limited studies are available in comparison with simplified (one- or two-step) chemistry, given the relatively large computational expense required and the current computational resources. A set of 8–9 chemical species and 18–21 elementary reactions are generally employed for H_2 systems [72, 112], while 21–22 species and 34–38 reactions are used for simple hydrocarbons systems [114]. These studies highlighted that the use of detailed chemistry is needed to accurately predict the energy-release pattern in RDEs and complex characteristics, including re-ignition, number of triple points and transverse waves [68].

5.5 Linearised model detonation engines

A linearised model may be constructed to simulate the operation of an RDE [79, 124]. These models, shown in **Figure 10**, are known as linearised model detonation engines (LMDEs). In this model, fuel is fed into the chamber, and a transverse

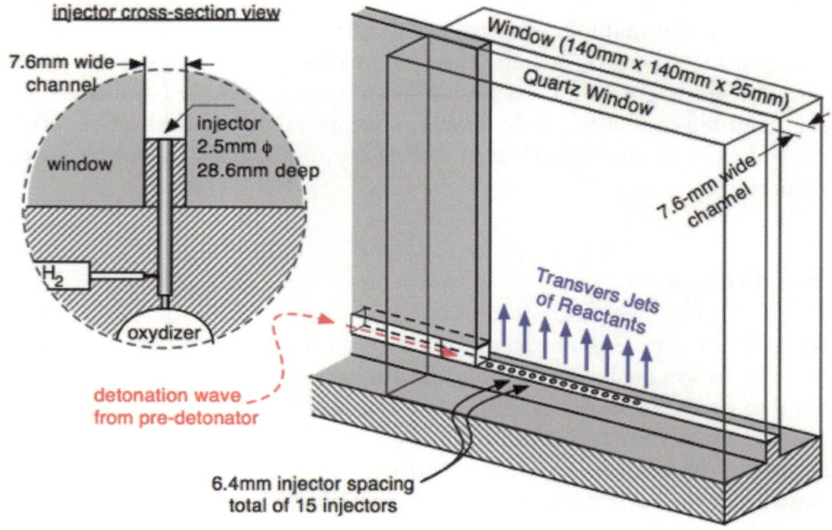

Figure 10.
An example linearised model detonation engine [79].

shock wave propagates through it. This occurs in much the same manner as in an RDE. However, the chamber is rectangular, and so the detonation only makes a single pass through the chamber [79, 124]. Both computer models and practical experiments have been run in three different modes, all using fresh supplies [79, 125]:

- The chamber is pre-filled with premixed fuel/oxidiser, and then the detonation is initiated.
- The chamber is pre-filled with an inert gas, then premixed fuel/oxidiser is injected and the detonation is initiated simultaneously.
- The chamber is pre-filled with oxidiser, then fuel is injected and the detonation is initiated simultaneously.

LMDEs have been used to characterise the detonation process, by allowing both sides of the chamber to be imaged through quartz walls, or the density field imaged through the use of the Schlieren technique [79, 126]. It has been found that the critical fill height of an LMDE is about 10λ, which is consistent with Eq. (4) for RDEs [27, 126]. It has been found that the presence of background gases, such as the inert gas used to pre-fill the chamber, strongly affected the detonation process, causing the reaction zone to slightly trail the detonation wave [125]. This produced fluctuations in the wave velocity, adversely affecting the detonation propagation [125]. This would seem to be consistent with mixing of detonated and undetonated reactants producing Kelvin-Helmholtz instabilities in an RDE, as noted in Section 3.1 [3, 22, 72, 73]. It was also found that low pressure zones in an LMDE attenuate reflected shocks [124]. This suggests that, should a shock wave be reflected off an irregular feature in an RDE's annulus, then the shock would not serve as a significant source of thermodynamic loss [124].

Computer modelling of an LMDE indicated that the propagation of a detonation wave was not affected by the turbulence caused by in-chamber mixing of fuel and oxidiser [118]. However, the presence of this turbulence did cause the reaction zone to trail the detonation wave [118]. A model of an LMDE was also used to test the result of applying different back pressures, such as might occur if a nozzle or a turbine was attached to an RDE [114]. This indicated that increased back pressure also increased the detonability of the fuel mixture, but also restricted the acceleration of the products, which, in some cases, led to the production of tertiary shock waves to sufficiently compress the flow to match the exit plane conditions [114]. However, as noted previously in Section 2, nozzles have very limited benefit [53], and, as noted in Section 4 the effect of secondary and tertiary shocks on a turbine may be problem.

6. Future outlook

Rotating detonation engines have the potential to provide a significantly more efficient combustion cycle than deflagration-based engines. The application of this technology to turbines promises to increase the thermodynamic efficiency of these engines to previously unattainable levels. Additionally, RDEs as a standalone engine hold significant promise for both air-breathing and air-independent rocket propulsion. However, there exists a large body of research and development work still-to-be undertaken, including:

- Nozzles have been shown to have limited benefit to the thrust generated by RDEs. However, varying the angles of the walls of an RDE, either independently or together, may simulate the effect of a nozzle to provide a slight benefit to performance. It remains unknown what effect such modifications to the conventional cylinder might have.

- Comparisons of thrust to weight ratios between experimental RDEs and conventional rocket engines show similar values, indicating that an RDE could represent a method of propulsion in space. This has not been widely explored as an option, and would benefit from experimental work in vacuum conditions or microgravity conditions.

- It has been suggested that there may be a maximum equivalence ratio at which an RDE will operate, but further investigation is required to determine if this is a universal limit, and identify ways to lower the limit.

- Triple points appear to have significant effect on the propagation of the detonation wave but little work has been done on determining the constraints, besides chemical composition, on the formation of stable and consistent triple points as well as the effect of those parameters on other characteristics of the triple points such as peak pressure and propagation direction. Findings would be beneficial in terms of properly defining the parameters that affect λ as well.

- Very few studies have provided a mathematical relationship between the detonation cell width and the geometry requirements of the chamber. More supporting work to help refine and verify or dispute the relationships that have been established needs to be done, so that in the future, specialised design needs can be catered for through knowing the geometry and cell width of fuel types.

- Varying the channel width has been noted to affect the stability of the detonation wave in an RDE. As such, this is likely to affect the performance of such devices. Further research is required to determine what the optimal width would be for different design requirements.

- It is established that RDE chambers need to be at least twice as long as the fuel fill height, and increasing the length four to six times the fill height improves the efficiency. However, depending on the ratio of inlet pressure to nozzle pressure, such a length increase may also result in reduced I_{sp}. Further research is required to determine an appropriate balance of these effects, and the effect chamber length has on other design parameters.

- So-called "centrebodiless" designs have been explored, and proposed for use in afterburners. However, they have not been modelled or tested with heated high velocity air, as would be typically found at the outlet of a conventional jet engine, so their potential performance remains unknown.

- It has been demonstrated that the thrust produced by RDEs scales non-linearly with engine size, but they are not yet approaching the size required to replace most existing gas turbines. It remains unknown if an RDE can be scaled up sufficiently to provide the thrust levels offered by contemporary gas turbine engines.

- It has been suggested that a turbine could be attached to an RDE. However, the effects of the various shocks on a turbine have not been explored. In particular, the oblique shock (Feature B in **Figure 5a**) has been shown to propagate out of the chamber, and is likely to have significant effect on the viability of using a turbine.

- The invsicid Euler equations have been demonstrated to over-predict deflagration in three-dimensional computational models of premixed RDEs, even with the use of detailed chemistry. Their validity in non-premixed RDE configurations, with deflagration upstream of the detonation and the potential to produce lifted detonation waves, still requires rigorous assessment.

- Viscous and thermal wall-effects in RDEs have significant effect on RDE performance characteristics, and may be essential in accurately reproducing experimentally measured values. Understanding of the appropriate numerical modelling approaches of these effects, however, is still immature, owing to the computational resources required for sufficiently fine resolution of near-wall grids.

- The computationally predicted wave-speeds and plenum pressures in RDEs are significantly different to those measured experimentally. It has been proposed that this could be partially due to baroclinic vorticity, resulting from interactions between detonation waves, fresh reactants, deflagration reaction-zones and post-combustion products, although this is yet to be analysed in detail in either full RDEs or linearised models.

Author details

Ian J. Shaw[1†], Jordan A.C. Kildare[1†], Michael J. Evans[1], Alfonso Chinnici[1], Ciaran A.M. Sparks[1], Shekh N.H. Rubaiyat[2], Rey C. Chin[1] and Paul R. Medwell[1*]

1 The University of Adelaide, Australia

2 University of South Australia, Australia

*Address all correspondence to: paul.medwell@adelaide.edu.au

† Joint first authors.

References

[1] Cullen R, Nicholls J, Ragland K. Feasibility studies of a rotating detonation wave rocket motor. Journal of Spacecraft and Rockets. 1966;**3**(6): 893-898

[2] Eidelman S, Grossman W, Lottati I. Review of propulsion applications and numerical simulations of the pulsed detonation engine concept. Journal of Propulsion and Power. 1991;**7**(6): 857-865

[3] Lu FK, Braun EM. Rotating detonation wave propulsion: Experimental challenges, modeling, and engine concepts. Journal of Propulsion and Power. 2014;**30**(5):1125-1142

[4] Bykovskii FA, Zhdan SA, Vedernikov EF. Continuous spin detonation of hydrogen-oxygen mixtures. 2. Combustor with an expanding annular channel. Combustion, Explosion, and Shock Waves. 2008;**44**(3):330-342

[5] Frolov S, Aksenov V, Ivanov V, Shamshin I. Large-scale hydrogenair continuous detonation combustor. International Journal of Hydrogen Energy. 2015;**40**(3):1616-1623

[6] Schwer D, Kailasanath K. Numerical investigation of the physics of rotating-detonation-engines. Proceedings of the Combustion Institute. 2011;**33**(2): 2195-2202

[7] Kailasanath K. Review of propulsion applications of detonation waves. AIAA Journal. 2000;**38**(9):1698-1708

[8] Heiser WH, Pratt DT. Thermodynamic cycle analysis of pulse detonation engines. Journal of Propulsion and Power. 2002;**18**(1):68-76

[9] Wintenberger E, Shepherd J. Thermodynamic cycle analysis for propagating detonations. Journal of Propulsion and Power. 2006;**22**(3): 694-698

[10] Yi T-H, Lou J, Turangan C, Choi J-Y, Wolański P. Propulsive performance of a continuously rotating detonation engine. Journal of Propulsion and Power. 2011;**27**(1):171-181

[11] Rotaru C, Mihăilă-Andres M, Renard P. New constructive solutions for aircraft turbojet engine combustion chamber. Journal of Military Technology—MTA Review. 2013;**23**(4): 219-230

[12] Nicholls JA, Wilkinson HR, Morrison RB. Intermittent detonation as a thrust-producing mechanism. Journal of Jet Propulsion. 1957;**27**(5):534-541

[13] Dunlap R. A preliminary study of the application of steady-state detonative combustion to a reaction engine. Journal of Jet Propulsion. 1958;**28**(7):451-456

[14] Wu Y, Ma F, Yang V. System performance and thermodynamic cycle analysis of airbreathing pulse detonation engines. Journal of Propulsion and Power. 2003;**19**(4):556-567

[15] Kailasanath K. Recent developments in the research on pulse detonation engines. AIAA Journal. 2003;**41**(2): 145-159

[16] Bussing T, Pappas G. An introduction to pulse detonation engines. In: 32nd Aerospace Sciences Meeting and Exhibit. 1994. p. 263

[17] Eidelman S, Yang X. Analysis of the pulse detonation engine efficiency. In: 34th AIAA/ASME/SAE/ASEE Joint Propulsion Conference & Exhibit. 1998. p. 3877

[18] Voitsekhovskii BV. The spin detonation. Doklady Akademii Nauk. 1957;**114**(4):717-720

[19] Yetao S, Meng L, Jianping W. Continuous detonation engine and effects of different types of nozzle on its propulsion performance. Chinese Journal of Aeronautics. 2010;**23**(6): 647-652

[20] Yi T-H, Turangan C, Lou J, Wolański P, Kindracki J. A three-dimensional numerical study of rotational detonation in an annular chamber. In: 47th AIAA Aerospace Sciences Meeting including The New Horizons Forum and Aerospace Exposition. 2009. p. 634

[21] Nakayama H, Moriya T, Kasahara J, Matsuo A, Sasamoto Y, Funaki I. Stable detonation wave propagation in rectangular-cross-section curved channels. Combustion and Flame. 2012; **159**(2):859-869

[22] Katta VR, Cho KY, Hoke JL, Codoni JR, Schauer FR, Roquemore WM. Effect of increasing channel width on the structure of rotating detonation wave. Proceedings of the Combustion Institute. 2019;**37**(3): 3575-3583

[23] Lee S-Y, Watts J, Saretto S, Pal S, Conrad C, Woodward R, et al. Deflagration to detonation transition processes by turbulence-generating obstacles in pulse detonation engines. Journal of Propulsion and Power. 2004; **20**(6):1026-1036

[24] Eidelman S, Grossmann W. Pulsed detonation engine experimental and theoretical review. In: 28th Joint Propulsion Conference and Exhibit. 1992. p. 3168

[25] Sorin R, Zitoun R, Desbordes D. Optimization of the deflagration to detonation transition: Reduction of length and time of transition. Shock Waves. 2006;**15**(2):137-145

[26] Kindracki J, Wolański P, Gut Z. Experimental research on the rotating detonation in gaseous fuels–oxygen mixtures. Shock Waves. 2011;**21**(2): 75-84

[27] Bykovskii FA, Zhdan SA, Vedernikov EF. Continuous spin detonations. Journal of Propulsion and Power. 2006;**22**(6):1204-1216

[28] Bykovskii FA, Zhdan SA, Vedernikov EF. Continuous spin detonation of fuel-air mixtures. Combustion, Explosion, and Shock Waves. 2006;**42**(4):463-471

[29] Suchocki J, Yu S-T, Hoke J, Naples A, Schauer F, Russo R. Rotating detonation engine operation. In: 50th AIAA Aerospace Sciences Meeting including the New Horizons Forum and Aerospace Exposition. 2012. p. 119

[30] Russo RM. Operational characteristics of a rotating detonation engine using hydrogen and air [MA thesis]. Air Force Institute of Technology; 2011

[31] Wolański P. Detonative propulsion. Proceedings of the Combustion Institute. 2013;**34**(1):125-158

[32] Schwer D, Kailasanath K. Numerical investigation of rotating detonation engines. In: 46th AIAA/ASME/SAE/ASEE Joint Propulsion Conference & Exhibit. 2010. p. 6880

[33] Schauer F, Stutrud J, Bradley R. Detonation initiation studies and performance results for pulsed detonation engine applications. In: 39th Aerospace Sciences Meeting and Exhibit. 2001. p. 1129

[34] Helman D, Shreeve R, Eidelman S. Detonation pulse engine. In: 22nd Joint Propulsion Conference. 1986. p. 1683

[35] Braun EM, Lu FK, Wilson DR, Camberos JA. Airbreathing rotating detonation wave engine cycle analysis.

Aerospace Science and Technology. 2013;**27**(1):201-208

[36] Schwer D, Kailasanath K. Fluid dynamics of rotating detonation engines with hydrogen and hydrocarbon fuels. Proceedings of the Combustion Institute. 2013;**34**(2): 1991-1998

[37] Shank J, King P, Karnesky J, Schauer F, Hoke J. Development and testing of a modular rotating detonation engine. In: 50th AIAA Aerospace Sciences Meeting including the New Horizons Forum and Aerospace Exposition. 2012. p. 120

[38] Fotia M, Kaemming TA, Hoke J, Schauer F. Study of the experimental performance of a rotating detonation engine with nozzled exhaust flow. In: 53rd AIAA Aerospace Sciences Meeting. 2015. p. 631

[39] Rankin BA, Codoni JR, Cho KY, Hoke JL, Shauer FR. Investigation of the structure of detonation wave in a non-premixed hydrogen air rotating detonation engine using mid-infrared imaging. Proceedings of the Combustion Institute. 2018:3479-3486

[40] Shank JC. Development and testing of a rotating detonation engine run on hydrogen and air [MA thesis]. Air Force Institute of Technology; 2012

[41] Hishida M, Fujiwara T, Wolański P. Fundamentals of rotating detonations. Shock Waves. 2009;**19**(1):1-10

[42] Schwer D, Kailasanath K. Numerical study of the effects of engine size in rotating detonation engines. In: 49th AIAA Aerospace Sciences Meeting including the New Horizons Forum and Aerospace Exposition. 2011. p. 581

[43] Zhdan SA, Bykovskii FA, Vedernikov EF. Mathematical modeling of a rotating detonation wave in a hydrogen-oxygen mixture. Combustion, Explosion, and Shock Waves. 2007; **43**(4):449-459

[44] Nordeen CA, Schwer D, Schauer F, Hoke J, Barber T, Cetegen B. Thermodynamic model of a rotating detonation engine. Combustion, Explosion, and Shock Waves. 2014; **50**(5):568-577

[45] Zhou R, Wu D, Wang J. Progress of continuously rotating detonation engines. Chinese Journal of Aeronautics. 2016;**29**(1):15-29

[46] Wolanski P. Application of the continuous rotating detonation to gas turbine. Applied Mechanics and Materials. 2015;**782**:3-12

[47] Kindracki J, Kobiera A, Wolanski P. Experimental and numerical research on rotating detonation in a small rocket engine model. International Congress of Diesel Engines. 2009:392-400

[48] St. George AC, Driscoll RB, Munday DE, Gutmark EJ. Development of a rotating detonation engine facility at the university of cincinnati. In: 53rd AIAA Aerospace Sciences Meeting. 2015. p. 635

[49] Yang C, Wu X, Ma H, Peng L, Gao J. Experimental research on initiation characteristics of a rotating detonation engine. Experimental Thermal and Fluid Science. 2016;**71**:154-163

[50] Stoddard W, St. George AC, Driscoll RB, Anand V, Gutmark EJ. Experimental validation of expanded centerbodiless RDE design. In: 54th AIAA Aerospace Sciences Meeting. 2016. p. 128

[51] Kailasanath K. Research on pulse detonation combustion systems: A status report. In: 47th AIAA Aerospace Sciences Meeting including The New Horizons Forum and Aerospace Exposition. 2009. p. 631

[52] Rankin BA, Fotia ML, Naples AG, Stevens CA, Hoke JL, Kaemming TA, et al. Overview of performance, application, and analysis of rotating detonation engine technologies. Journal of Propulsion and Power. 2016;**33**:131-143

[53] Yi T-H, Lou J, Turangan C, Khoo BC, Wolański P. Effect of nozzle shapes on the performance of continuously-rotating detonation engine. In: 48th AIAA Aerospace Sciences Meeting Including the New Horizons Forum and Aerospace Exposition. 2010. p. 152

[54] Kato Y, Gawahara K, Matsuoka K, Kasahara J, Matsuo A, Funaki I, et al. Thrust measurement of rotating detonation engine by sled test. In: 50th AIAA/ASME/SAE/ASEE Joint Propulsion Conference. 2014. p. 4034

[55] Aircraft Design Group. Data on Large Turbofan Engines. Available from: http://adg.stanford.edu/aa241/propulsion/largefan.html. Leland Stanford Junior University. 2001. Available from: https://web.archive.org/web/20010223235152/http://adg.stanford.edu/aa241/propulsion/largefan.html

[56] Leyes RA, Fleming W. The History of North American Small Gas Turbine Aircraft Engines. AIAA; 1999. 998 pp

[57] Pratt & Whitney. F135 Engine. 2018. Available from: http://newsroom.pw.utc.com/download/me_f135_engine_cv_pcard.pdf

[58] General Electric Aviation. F414 Turbofan Engines. 2014. Available from: https://www.geaviation.com/sites/default/files/datasheet-F414-Family.pdf

[59] Niaz S, Manzoor T, Pandith AH. Hydrogen storage: Materials, methods and perspectives. Renewable and Sustainable Energy Reviews. 2015;**50**:457-469

[60] Liu S, Lin Z, Liu W, Lin W, Zhuang F. Experimental realization of H_2/air continuous rotating detonation in a cylindrical combustor. Combustion Science and Technology. 2012;**184**(9): 1302-1317

[61] Lin W, Zhou J, Liu S, Lin Z. An experimental study on CH_4/O_2 continuously rotating detonation wave in a hollow combustion chamber. Experimental Thermal and Fluid Science. 2015;**62**:122-130

[62] Kindracki J, Kobiera A, Wolański P, Gut Z, Folusiak M, Swiderski K. Experimental and numerical study of the rotating detonation engine in hydrogen-air mixtures. Progress in Propulsion Physics. 2011;**2**:555-582

[63] Naples A, Hoke J, Karnesky J, Schauer F. Flowfield characterization of a rotating detonation engine. In: 51st AIAA Aerospace Sciences Meeting including the New Horizons Forum and Aerospace Exposition. 2013. p. 278

[64] Rankin BA, Richardson DR, Caswell AW, Naples A, Hoke J, Schauer F. Imaging of OH* chemiluminescence in an optically accessible nonpremixed rotating detonation engine. In: 53rd AIAA Aerospace Sciences Meeting. 2015. p. 1604

[65] Cho KY, Codoni JR, Rankin BA, Hoke J, Schauer F. High-repetition-rate chemiluminescence imaging of a rotating detonation engine. In: 54th AIAA Aerospace Sciences Meeting. 2016. p. 1648

[66] Voitsekhovskii BV, Mitrofanov VV, Topchian ME. Investigation of the structure of detonation waves in gases. Symposium (International) on Combustion. 1969;**12**(1):829-837

[67] Cocks PA, Holley AT, Greene CB, Haas M. Development of a high fidelity RDE simulation capability. In: 53rd AIAA Aerospace Sciences Meeting. 2015. p. 1823

[68] Oran ES, Weber JW, Stefaniw EI, Lefebvre MH, Anderson JD. A numerical study of a two-dimensional H_2-O_2-Ar detonation using a detailed chemical reaction model. Combustion and Flame. 1998;**113**(1):147-163

[69] Lee JH. Dynamic parameters of gaseous detonations. Annual Review of Fluid Mechanics. 1984;**16**(1):311-336

[70] Kessler D, Gamezo V, Oran E. Gas-phase detonation propagation in mixture composition gradients. Philosophical Transactions of the Royal Society A: Mathematical, Physical and Engineering Sciences. 2012;**370**(1960): 567-596

[71] Fujiwara T, Hishida M, Kindracki J, Wolański P. Stabilization of detonation for any incoming mach numbers. Combustion, Explosion, and Shock Waves. 2009;**45**(5). Translation of article originally published in Russian in Fizika Goreniya i Vzryva, 2009;**45**(5): 108-110, 603-605

[72] Hayashi AK, Kimura Y, Yamada T, Yamada E, Kindracki J, Dzieminska E, et al. Sensitivity analysis of rotating detonation engine with a detailed reaction model. In: 47th AIAA Aerospace Sciences Meeting including The New Horizons Forum and Aerospace Exposition. 2009. p. 633

[73] Tsuboi N, Watanabe Y, Kojima T, Hayashi AK. Numerical estimation of the thrust performance on a rotating detonation engine for a hydrogen-oxygen mixture. Proceedings of the Combustion Institute. 2015;**35**(2): 2005-2013

[74] Sousa J, Paniagua G, Morata EC. Thermodynamic analysis of a gas turbine engine with a rotating detonation combustor. Applied Energy. 2017;**195**:247-256

[75] Dremin ANA. Toward detonation theory. In: Shock Wave and High Pressure Phenomena. 1999. pp. 1-14. ISBN 9781461205630

[76] Vasil'ev AA, Gavrilenko TP, Topchiyan ME. Chapman-Jouguet condition for real detonation waves. Combustion, Explosion and Shock Waves. 1973;**9**(2):268-272

[77] Eto K, Tsuboi N, Hayashi AK. Numerical study on three-dimensional CJ detonation waves: Detailed propagating mechanism and existence of OH radical. Proceedings of the Combustion Institute. 2005;**30**(2): 1907-1913

[78] Yu J, Hou B, Lelyakin A, Xu Z, Jordan T. Gas detonation cell width prediction model based on support vector regression. Nuclear Engineering and Technology. 2017;**49**(7):1423-1430

[79] Prakash S, Fiévet R, Raman R, Burr JR, Yu KH. Numerical study of the detonation wave structure in a linear model detonation engine. In: 2018 Joint Propulsion Conference. 2018. p. 4966

[80] St. George AC, Driscoll RB, Anand V, Munday DE, Gutmark EJ. Fuel blending as a means to achieve initiation in a rotating detonation engine. In: 53rd AIAA Aerospace Sciences Meeting. 2015. p. 633

[81] Yao S, Wang J. Numerical investigation of effects of fuel injection on rotating detonation engine. In: 51st AIAA/SAE/ASEE Joint Propulsion Conference. 2015. p. 4192

[82] Dyer R, Naples A, Kaemming T, Hoke J, Schauer F. Parametric testing of a unique rotating detonation engine design. In: 50th AIAA Aerospace Sciences Meeting including the New Horizons Forum and Aerospace Exposition. 2012. p. 121

[83] Naples A, Hoke J, Schauer F. Rotating detonation engine interaction with an annular ejector. In: 52nd

Aerospace Sciences Meeting. 2014. p. 287

[84] Liu Y, Wang Y, Li Y, Li Y, Wang J. Spectral analysis and self-adjusting mechanism for oscillation phenomenon in hydrogen-oxygen continuously rotating detonation engine. Chinese Journal of Aeronautics. 2015;**28**(3): 669-675

[85] Liu S-J, Lin Z-Y, Liu W-D, Lin W, Sun M-B. Experimental and three-dimensional numerical investigations on H_2/air continuous rotating detonation wave. Proceedings of the Institution of Mechanical Engineers, Part G: Journal of Aerospace Engineering. 2012;**227**(2): 326-341

[86] Anand V, George AS, Driscoll R, Gutmark E. Characterization of instabilities in a rotating detonation combustor. International Journal of Hydrogen Energy. 2015;**40**(46): 16649-16659

[87] Zhou R, Wang J-P. Numerical investigation of shock wave reflections near the head ends of rotating detonation engines. Shock Waves. 2013; **23**(5):461-472

[88] Fotia M, Hoke J, Schauer F. Experimental ignition characteristics of a rotating detonation engine under backpressured conditions. In: 53rd AIAA Aerospace Sciences Meeting. 2015. p. 632

[89] Meng Q, Zhao N, Zheng H, Yang J, Qi L. Numerical investigation of the effect of inlet mass flow rates on H_2/air non-premixed rotating detonation wave. International Journal of Hydrogen Energy. 2018;**43**(29):13618-13631

[90] Rankin BA, Richardson DR, Caswell AW, Naples AG, Hoke JL, Schauer FR. Chemiluminescence imaging of an optically accessible non-premixed rotating detonation engine. Combustion and Flame. 2017;**176**:12-22

[91] Wolański P. Rotating detonation wave stability. In: 23rd International Colloquium on the Dynamics of Explosions and Reactive Systems. 2011. pp. 1-6

[92] Falempin F, Daniau E. A contribution to the development of actual continuous detonation wave engine. In: 15th AIAA International Space Planes and Hypersonic Systems and Technologies Conference. 2008. p. 2679

[93] Stechmann DP. Experimental study of high-pressure rotating detonation combustion in rocket environments English [PhD thesis]. Purdue University; 2017

[94] Cocks PA, Holley AT, Rankin BA. High fidelity simulations of a non-premixed rotating detonation engine. In: 54th AIAA Aerospace Sciences Meeting. 2016. p. 125

[95] Shao Y-T, Liu M, Wang J-P. Numerical investigation of rotating detonation engine propulsive performance. Combustion Science and Technology. 2010;**182**(11–12):1586-1597

[96] Davidenko D, Gökalp I, Kudryavtsev A. Numerical study of the continuous detonation wave rocket engine. In: 15th AIAA International Space Planes and Hypersonic Systems and Technologies Conference. 2008. p. 2680

[97] Liu S-J, Lin Z-Y, Sun M-B, Liu W-D. Thrust vectoring of a continuous rotating detonation engine by changing the local injection pressure. Chinese Physics Letters. 2011;**28**(9):1-4

[98] Schwer D, Kailasanath K. Effect of inlet on fill region and performance of rotating detonation engines. In: 47th AIAA/ASME/SAE/ASEE Joint Propulsion Conference & Exhibit. 2011. p. 6044

[99] Schwer D, Kailasanath K. Feedback into mixture plenums in rotating detonation engines. In: 50th AIAA Aerospace Sciences Meeting including the New Horizons Forum and Aerospace Exposition. 2012. p. 617

[100] Chang P-H, Leong WK, Li J-M, Teo CJ, Khoo BC. Investigation of channel pressure effect on rotating detonation engine. In: AIAA Scitech 2019 Forum. 2019. p. 2020

[101] Zhou R, Wang J-P. Numerical investigation of flow particle paths and thermodynamic performance of continuously rotating detonation engines. Combustion and Flame. 2012; **159**(12):3632-3645

[102] Massa L, Chauhan M, Lu F. Detonation-turbulence interaction. Combustion and Flame. 2011;**158**(9): 1788-1806

[103] Driscoll R, Anand V, St George A, Gutmark E. Investigation on RDE operation by geometric variation of the combustor annulus and nozzle exit area. In: 9th US National Combustion Meeting. 2015. pp. 1-10

[104] Bussing T. A rotary valved multiple pulse detonation engine. In: 31st Joint Propulsion Conference and Exhibit. 1995. p. 2577

[105] Gavrikov A, Efimenko A, Dorofeev S. A model for detonation cell size prediction from chemical kinetics. Combustion and Flame. 2000;**120**(1–2): 19-33

[106] Pfahl U, Schultz E, Shepherd J. Detonation cell width measurements for H_2–N_2O–N_2–O_2–CH_4–NH_3 mixtures. Technical Report. 1998

[107] Frolov S, Dubrovskii A, Ivanov V. Three-dimensional numerical simulation of operation process in rotating detonation engine. Progress in Propulsion Physics. 2013;**4**:467-488

[108] Paxson DE. Numerical analysis of a rotating detonation engine in the relative reference frame. In: 52nd Aerospace Sciences Meeting. 2014. p. 284

[109] Nordeen CA, Schwer D, Schauer F, Hoke J, Barber T, Cetegen BM. Role of inlet reactant mixedness on the thermodynamic performance of a rotating detonation engine. Shock Waves. 2016;**26**(4):417-428

[110] Wu D, Zhou R, Liu M, Wang J. Numerical investigation of the stability of rotating detonation engines. Combustion Science and Technology. 2014;**186**(10–11):1699-1715

[111] Davidenko DM, Jouot F, Kudryavtsev AN, Dupré G, Gökalp I, Daniau E, et al. Continuous detonation wave engine studies for space application. Progress in Propulsion Physics. 2009;**1**:353-266

[112] Uemura Y, Hayashi AK, Asahara M, Tsuboi N, Yamada E. Transverse wave generation mechanism in rotating detonation. Proceedings of the Combustion Institute. 2013;**34**(2): 1981-1989

[113] Driscoll R, George AS, Gutmark EJ. Numerical investigation of injection within an axisymmetric rotating detonation engine. International Journal of Hydrogen Energy. 2016;**41**(3): 2052-2063

[114] Sato T, Voelkel S, Raman V. Analysis of detonation structures with hydrocarbon fuels for application towards rotating detonation engines. In: 2018 Joint Propulsion Conference. 2018. p. 4965

[115] Bykovskii FA, Vedernikov EF. Continuous detonation of a subsonic flow of a propellant. Combustion, Explosion and Shock Waves. 2003; **39**(3):323-334

[116] Lu T, Law CK, Ju Y. Some aspects of chemical kinetics in chapman-Jouguet detonation: Induction length analysis. Journal of Propulsion and Power. 2003;**19**(5):901-907

[117] Westbrook CK. Hydrogen oxidation kinetics in gaseous detonations. Combustion Science and Technology. 1982;**29**(1–2):67-81

[118] Masselot D, Fiévet R, Raman V. Effect of equivalence ratio and turbulence fluctuations on the propagation of detonations. In: 55th AIAA Aerospace Sciences Meeting. 2017. p. 374

[119] Roy A, Strakey P, Sidwell T, Ferguson DH. Unsteady heat transfer analysis to predict combustor wall temperature in rotating detonation engine. In: 51st AIAA/SAE/ASEE Joint Propulsion Conference. 2015. p. 902

[120] Strakey P, Ferguson D, Sisler A, Nix A. Computationally quantifying loss mechanisms in a rotating detonation engine. In: 54th AIAA Aerospace Sciences Meeting. 2016. p. 900

[121] Falempin F, Daniau E, Getin N, Bykovskii F, Zhdan S. Toward a continuous detonation wave rocket engine demonstrator. In: 14th AIAA/AHI Space Planes and Hypersonic Systems and Technologies Conference. 2006. p. 7956

[122] Asahara M, Tsuboi N, Hayashi AK, Yamada E. Two-dimensional simulation on propagation mechanism of H_2/O_2 cylindrical detonation with a detailed reaction model: Influence of initial energy and propagation mechanism. Combustion Science and Technology. 2010;**182**(11–12):1884-1900

[123] Pope S. PDF methods for turbulent reactive flows. Progress in Energy and Combustion Science. 1985;**11**(2):119-192

[124] Burr JR, Yu KH. Shock in reactive cross-flow under partial confinement. In: International Colloquium on the Dynamics of Explosions and Reactive Systems. 2015. pp. 1-6

[125] Burr JR, Yu K. Detonation wave propagation in cross-flow of discretely spaced reactant jets. In: 53rd AIAA/SAE/ASEE Joint Propulsion Conference. 2017. p. 4908

[126] Burr JR, Yu KH. Blast wave propagation in cross-flow of detonable mixture. In: 50th AIAA/ASME/SAE/ASEE Joint Propulsion Conference. 2014. p. 3984

Section 3

Special Topics

Chapter 8

Effect of Laminar Flow on the Corrosion Activity of AA6061-T6 in Seawater

Gloria Acosta, Lucien Veleva, Luis Chávez and Juan L. López

Abstract

The electrochemical behaviour and surface changes on AA6061-T6 alloy exposed to Caribbean seawater from the Cozumel Channel for 30 days under laminar flow (0.1 ms^{-1}) were studied, these contrasting then with stationary conditions. Open circuit potential monitoring and electrochemical current fluctuations, considered as electrochemical noise (EN), were employed as two nondestructive methods. The calculated corrosion current, based on Rn, was one order higher in laminar flow. The fluctuations of current were transformed in the frequency domain. Their power spectral density (PSD) plots were obtained in order to gain information concerning the dynamic of the spontaneous release of energy during the corrosion process. The value of the exponent β in PSD graphs suggested that the localised corrosion on AA6061-T6 surface occurs as a persistent stationary process, which dynamic is controlled by oxygen diffusion. The changes in the morphology and elemental composition of the formed layers revealed that the localised attacks occurred in the vicinity of intermetallic particles rich in Fe and Cu, which act as cathodes.

Keywords: aluminium alloy 6061-T6, seawater, laminar flow, intermetallic particles, electrochemical current noise

1. Introduction

Aluminium alloy (AA) 6061-T6 is popular as a nonferrous material for structures in seawater [1] and is characterised by various properties, such as strength-to-weight ratio, extrudability—particularly for the manufacture of profiles with complex geometry, low thermal expansion coefficient, good wear resistance and corrosion resistance [2]. The alloy presents good corrosion resistance in many environments having neutral pH, because of the formation of protective amorphous aluminium oxide film on its surface of approximately 2–3 nm thickness and is insoluble in water [3, 4]. The addition of alloying elements to aluminium increases its mechanical properties [5]; however, the precipitated intermetallic particles (IMPs) have a harmful effect on the corrosion resistance of the Al alloys [4, 6–10]. IMPs present electrochemical behaviour different from the alloy matrix, and they may be classified in two types: cathodic and anodic [7, 9–12]. The electrochemically active anodic particles are rich in Mg, Si and Al, with Mg preferential dissolution, leaving a cavity in the oxide layer, while the cathodic particles are rich in Fe, Si and Cr, acting as preferential sites for oxygen reduction.

The localised corrosion becomes greater in the presence of aggressive ions, such as chlorides [4, 13–15], and corrosion pits initiate in oxide film sites, weakened by chloride attack. Moreover, the heterogeneity of the surface could result in favourable nucleation sites.

Seawater is a complex electrolyte of different ions, with a high salinity (3.5%, density 1.023 g/cm^3 at 25°C), which causes damage to metals in a short time [16, 17]. The principal parameters that affect the corrosion behaviour of metals immersed in this electrolyte are oxygen content, dissolved mineral salts, pH, temperature, specific contaminants and flow velocity [16, 18].

The characterisation of the corrosion process requires electrochemical nondestructive techniques, and the preferred methods are those that do not apply external polarisation. The monitoring of the open circuit potential (φ_{corr}, free corrosion potential) or corrosion current is one of these and is easy to handle. Fluctuations may be interpreted as electrochemical noise (EN), which is useful for the purposes of corrosion mechanism characterisation. Electrochemical noise is presented by random fluctuations of corrosion potential or current, typically with frequencies below 10 Hz and low amplitude [19]. This technique can provide information concerning the nature of the corrosion process and the rate thereof. The main sources of EN observed in corrosion systems are attributed to microscopic and macroscopic events [19–21]. EN measurements can be performed under corrosion potential or any constant potential/current, depending on the research objective, to analyse the corrosion mechanism and obtain the corrosion rate [20].

EN measurements can be analysed transforming the data in the frequency domain by fast Fourier transform (FFT) to obtain power spectral density (PSD) [22, 23]. PSD plots display a slope, β exponent, which enables the differentiation between series with fractional Gaussian noise (*fGn*, β from -1 to 1) and fractional Brownian motion (fBm, β from 1 to 3). The *fGn* is a stationary process, and the fBm is nonstationary [24, 25]. The β exponent is a parameter correlated with the strength of persistence in a process [26]. In our previous studies, EN technique was carried out to characterise the first stages of corrosion in stationary seawater of copper [27], aluminium [28] and aluminium alloys [29], as well as the initial stages of AZ31B Mg alloy in simulated body fluid [30].

The object of this study is to investigate the electrochemical behaviour and surface changes on AA6061-T6 alloy, exposed to Caribbean seawater (Cozumel Channel) under laminar flow, contrasting these with stationary flow. Two nondestructive electrochemical methods were used to test the corrosion resistance of the alloy. The corrosion current and φ_{corr} were considered as EN and transformed in the frequency domain, in order to gain information on the dynamics of the spontaneous release of energy during the corrosion process. X-ray photoelectron spectroscopy (XPS) measurement was employed to analyse the composition of the formed corrosion layers, as well as SEM-EDS surface analysis. To the best of own knowledge, there is still no study on the initial stages of localised corrosion of AA6061-T6 alloy in laminar flow conditions.

2. Experimental

2.1 Materials

The nominal composition of AA6061-T6 (Metal Plastic Mexicali) was (wt%) 1.10% Mg, 0.5% Fe, 0.4% Si, 0.31% Cu, 0.19% Cr, 0.07% Zn and 0.05% Ti and the remainder Al. The seawater was extracted from the Caribbean Cozumel Channel, at 10 km offshore, to minimise the effect of human pollution, and a depth of 10 m. The seawater composition and physicochemical properties were as follows: salinity 36.4 g L^{-1}, chlorides 20.12 g L^{-1}, sulphates 2.82 g L^{-1}, nitrates 0.48 μmol/L, nitrites 0.18 μmol/L, ammonia 0.99 μmol/L, pH 7.3, dissolved oxygen 5.8 mg L^{-1} and conductivity 51.6 mS.

Samples of 1 cm^2 were embedded in Epofix resin as working electrodes for electrochemical tests. In addition, samples of 1 cm^2 were cleaned with ethanol and immersed in 100 mL of seawater (triplicated) under flow conditions, for 0, 5, 15 and 30 days, respectively. Before exposure, the specimens were ground with silicon carbide paper to 4000 grit, using distilled water as lubricant. They were washed with distillated water and dried in air.

2.2 Laminar flow of seawater

The speed range found in the Cozumel Channel indicates that the ocean current speed oscillates between 0.1 and 1 m s^{-1}. The tubing employed to connect the peristaltic pump (LAMBDA Laboratory Instrument) with the electrochemical cell has a diameter of 6 mm, and using the lower limit of the flow speed (0.1 m s^{-1}), the calculated Reynolds number was 622.97, which corresponds to laminar hydrodynamic flow [31]. The dimensionless N_{Re} characterises the diffusion-dependent system and is governed by the ratio of the inertial forces acting on the fluid to the viscous forces:

$$N_{Re} = \frac{\rho \bar{u} d}{\mu} \tag{1}$$

where ρ is the fluid density (seawater, 997.8 kg m^{-3}), $\bar{u}$ is the flow speed (0.1 m s^{-1}), d is the tubing diameter (6 mm) and μ is the seawater viscosity (0.961×10^{-3} N s m^{-2}). All experiments were carried out at 21°C.

2.3 Surface analysis

AA6061-T6 samples were characterised before and after exposure in seawater by SEM-EDS (Philips-XL and ESEM-JEOL JSM-7600F), in order to observe the microstructure and elemental changes on their surface. Corrosion products formed on AA6061-T6 sample surfaces were identified through X-ray photoelectron spectroscopy (XPS, K-alpha, Thermo Scientific, Waltham, MA, USA). In addition, corrosion products were removed in accordance with ASTM G1-90 standard [32], and the alloy surfaces were reexamined.

2.4 Electrochemical tests

A multiport electrochemical cell kit (Gamry Instruments, 1 L total volume) and a Gamry PCI4G750-52103 potentiostat were used for all electrochemical measurements. The experimental setup employed was according to ASTM G199-09 standard [33]: two identical AA6061-T6 working electrodes (WEs) connected to a zero resistance ammeter (ZRA) and a saturated calomel electrode (SCE, Hg^{2+}/Hg_2Cl_2 = 0.244 V) as reference electrode. The electrodes were immersed in an electrochemical cell with seawater, employed as test electrolyte. All tests were carried out in laminar and stationary flow conditions up to 30 days. EN data were collected at different times during 3 h, initial, 1, 5, 15 and 30 days in OCP, with a sampling frequency of 10 Hz. The obtained data of current and potential fluctuations were plotted vs. time. The current fluctuations were considered as EN and processed in the frequency domain by fast FFT to graph PSD. The PSD as a function of low frequencies was analysed on bi-logarithmic scale (10^{-3} –1 Hz) of power per unit frequency (A^2/Hz) vs. frequency (Hz). These procedures allowed the fitting of a straight line and obtention of the β slope. The β value characterises the corrosion mechanism on AA6061-T6 surface. The processing of data was realised with Electrochemical Signal Analyser V.7.0.1 software (Gamry Instruments, Philadelphia, PA, USA). All measurements were checked in triplicate.

2.4.1 Energy spectral density

The energy spectral density expresses how the energy of a time series is dispersed with a frequency; it can reflect the change in system dynamic. For a signal X(t) the energy EE is [34]

$$E = \int_{-\infty}^{\infty} |x(t)|^2 dt \tag{2}$$

For signals with a finite total energy, an equivalent expression for the energy is expressed as

$$E = \int_{-\infty}^{\infty} |x(t)|^2 dt = \int_{-\infty}^{\infty} |\hat{x}(f)|^2 df \tag{3}$$

And using the Fourier transform of the signal:

$$\hat{x}(f) = \int_{-\infty}^{\infty} e^{-2\pi i f t} x(t) df \tag{4}$$

where f is the frequency in Hz.

The integral on the right-hand side of Eq. (3) is the energy of the signal, and the integrand $|\hat{x}(f)|^2$ describes the energy per unit frequency contained in the signal (ESD).

3. Results and discussion

3.1 Surface analysis

Figure 1 shows SEM images of the as-received specimen of 6061-T6 aluminium alloy surface. Small precipitates (labelled as A, B, C) may be seen (**Figure 1b**). In accordance with EDS, presented in **Table 1**, these precipitates correspond to particles rich in Fe and could be considered as elements of the following phases: Al_3Fe, Al–Si–Mn–Fe and α-Al(Fe, Mn, Cu) [9, 10, 35]. The elemental composition of the surface layer is indicated by zone D.

Figure 2a presents the SEM image of the film formed on the surface of the sample after 5 days of immersion in seawater from the Cozumel Channel under laminar flow. It can be clearly seen that the passive layer begins to break down around the alloying elements (**Figure 2b**). Probably, this layer is less protective in the vicinity of the intermetallic particles, causing the formation of local electrochemical cells with the Al matrix. EDS analysis, **Table 2**, confirmed the presence of particles mentioned above. According to EDS (**Table 2**), zone A is characterised by its higher content of Al, O and Cl, which could relate to corrosion products. According to previous studies, at pH > 8.5, $Al(H_2O)_6^{3+}$ cations appear, while in the range of pH 4.5–8.5, $Al(OH)_3$ predominates [36, 37]. In chloride solutions,

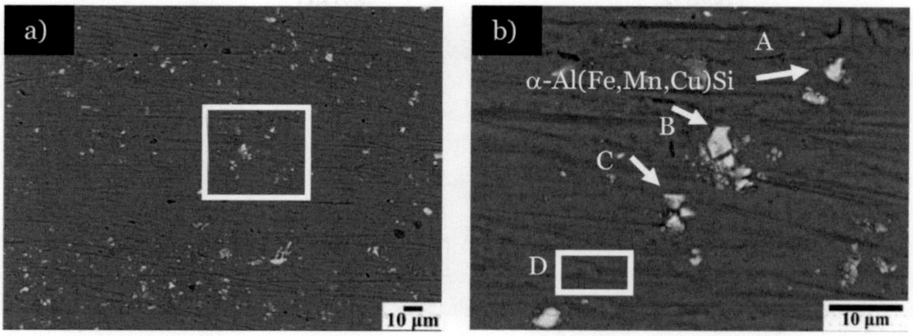

Figure 1.
(a) SEM images (×500) of AA6061-T6 surface before immersion in Caribbean seawater and (b) zoom zone (×2000).

Element	C	O	Mg	Al	Si	Cr	Mn	Fe	Cu
A	2.4	3.3	—	60.3	6.5	1.3	1.6	23.7	0.9
B	2.3	2.7	—	59.3	7.4	1.3	1.5	24.2	1.3
C	2.2	4.5	—	71.7	4.8	0.6	0.9	14.6	0.7
D	2.0	2.1	1.7	93.7	0.5	—	—	—	—

Table 1.
EDS analysis (wt%) of reference aluminium alloy 6061-T6 surface.

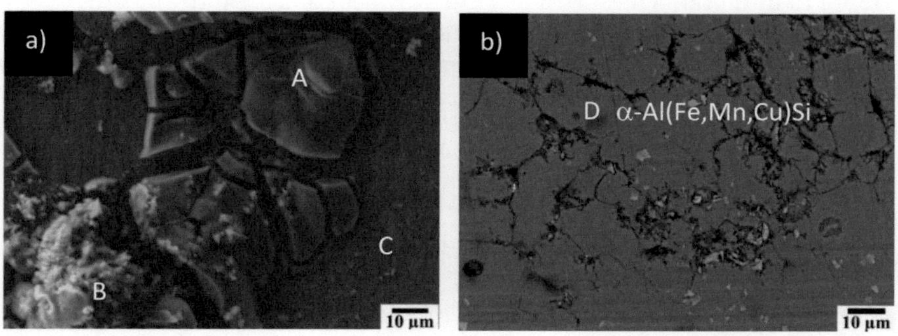

Figure 2.
SEM images of AA6061-T6 surface after exposure in Caribbean seawater under laminar flow for 5 days (a) and (b) after removal of corrosion layer.

					(a)					
Element	**C**	**O**	**Na**	**Mg**	**Al**	**Si**	**S**	**Cl**	**K**	**Ca**
A	—	47.4	0.7	1.2	25.1	—	1.8	23.3	—	0.5
B	6.4	56.6	1.9	1.0	29.2	—	0.4	0.5	0.5	3.5
C	4.4	11.5	—	0.8	81.2	0.4	0.4	0.3	0.8	0.2
					(b)					
Element	**C**	**O**	**Mg**	**Al**	**Si**	**Cr**	**Mn**	**Fe**	**Cu**	
D	10.9	4.7	0.4	58.0	5.8	0.6	1.2	17.6	0.8	

Table 2.
(a) EDS analysis (wt.%) of aluminium alloy 6061-T6 surface after 5 days of exposure in Caribbean seawater under laminar flow at 21°C and (b) after removal of corrosion products.

aluminium metal ionises rapidly to the Al^{3+} ion, which also hydrolyses very rapidly (owing to the negative potential value) [38]. Both of these Al cations can react with chloride ions and form $AlCl_3$ soluble in water (31.77 wt%) [39]; this is converted later to a relatively stable species of basic aluminium chloride ($AlCl_3$•H_2O), transformed slowly to $Al(OH)_3$ and finally to Al_2O_3•H_2O, an important corrosion product for the repassivation process of the aluminium surface [38].

Based on EDS analysis, in zone B in addition to oxygen and aluminium, elements of calcium and carbon were present, both possibly as a part of a $CaCO_3$ precipitate, originating from seawater. Meanwhile, the layer in zone C maintained a composition similar to that of the alloy, which indicates that the corrosion process was still beginning on the surfaces of similar areas.

After removal of the layer of corrosion products (**Figure 2b**), area damaged by pitting and cracking was observed on the alloy surface. However, some precipitates remained on the surface of AA6061-T6, which according to EDS (**Table 2**) correspond to cathodic particles rich in Fe, α-Al(Fe,Mn,Cu)Si, which promoted the preferential dissolution of the aluminium matrix (local alkalisation) [10].

Figure 3 compares SEM images of the aluminium alloy 6061-T6 after exposure at 30 days in laminar flow (**Figure 3a** and **c**) and stationary flow (**Figure 3b** and **d**). In these micrographs, the products formed on the surfaces of the alloy (**Figure 3a** and **b**) can be seen and compared, whose EDS analysis is summarised in **Table 3**. In laminar flow (**Figure 3a**), the segregation of particles rich in Cu (particles A) with the presence of O and Al could be considered as the phases of Al_2Cu, AlMgSiCu (Q-phase) and Al_7Cu_2Fe; both relatively resistant to corrosion, because they are nobler than the aluminium matrix [4, 12]. Q-phase acts as a cathode and does not contribute to intergranular corrosion when it is not connected to any other Cu particle. Otherwise, the Q-phase as cathode promotes the development of intergranular corrosion, initiated in the presence of chloride ions (NaCl). In the corrosion layer, particles without Cu appeared (particles B), rich in Fe, Al and Mg,

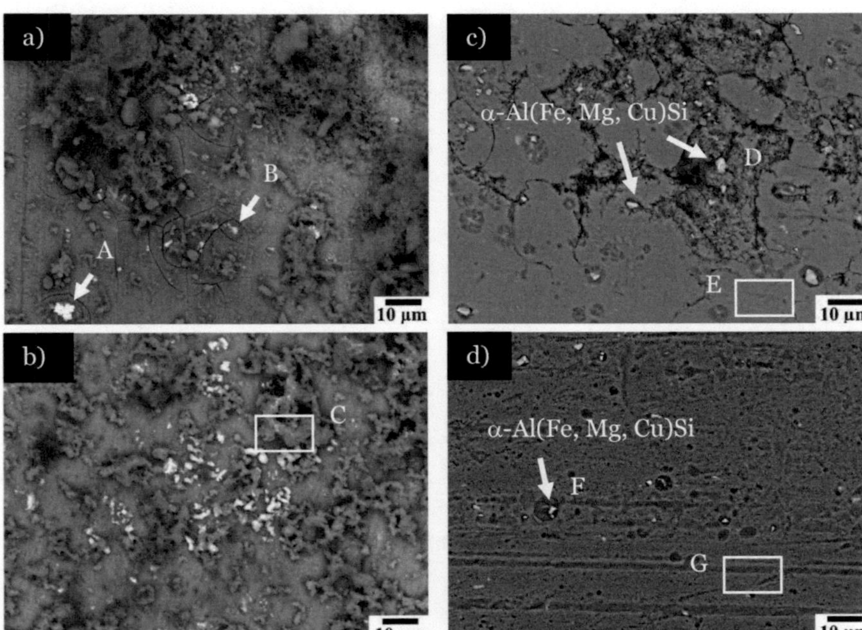

Figure 3.
SEM images of AA6061-T6 surface after 30 days of exposure in Caribbean seawater, (a) laminar and (b) stationary flow conditions; after removal of corrosion products, (c) laminar and (d) stationary flow conditions.

						(a)						
Element	**C**	**O**	**Na**	**Mg**	**Al**	**Si**	**Cr**	**Mn**	**Cl**	**Ca**	**Fe**	**Cu**
A	3.3	22.3	2.2	3.3	10.2	0.7	—	—	1.0	0.3	1.0	55.7
B	4.2	31.3	1.7	1.8	39.6	3.9	0.5	0.8	0.8	0.5	14.2	0.7
C	12.7	44.9	0.5	3.0	37.8	0.1	—	—	0.5	0.5	—	—
						(b)						
D	4.1	—	—	—	63.3	6.0	1.0	1.2	1.5	—	21.8	1.1
E	5.7	4.4	—	0.5	84.6	3.2	0.7	0.9	—	—	—	—
F	5.7	1.5	—	—	69.6	5.1	1.3	1.0	—	—	15.0	0.8
G	5.8	4.6	—	0.7	88.2	0.4	0.2	—	—	—	—	—

Table 3.
EDS analysis (wt%) of aluminium alloy 6061-T6 surfaces: (a) after 30 days of exposure in seawater under laminar flow and (b) after removal the layer of corrosion products.

which could be attributed to the following phases, Al–Si–Mn–Fe and Al–Mg–Si, reported for aluminium alloy series 6xxx [35]. Zone C of the layer formed under stationary flow (**Figure 3b**) presented a similar composition of the alloy (**Table 3**), however, with the oxide layer of $Al_2O_3{\bullet}H_2O$ on the alloy surface 6061-T6, as a transformation product of basic aluminium ($AlCl_3{\bullet}H_2O$) in the presence of NaCl [38]. This layer is part of the entire surface (**Figure 3a**), since the three zones (A, B and C) have a high oxygen content (**Table 3**) [4].

Figure 3c corresponds to the SEM image of the aluminium alloy 6061-T6 exposed in seawater at 30 days with flow, after removal of the layer of corrosion products. An area on the surface damaged by cracking and fissures can be observed, while the surface exposed to seawater without flow (**Figure 3d**) shows less damage, with pitting holes of several diameters. In contrast, on the surface exposed to laminar flow for this same time period, the pits are not clearly observable, but localised corrosion damage appeared in the form of cracks and fissures. This effect is due to the seawater flow, which accelerates the corrosion process, causing detachment of the destroyed passive layer and the appearance of new "fresh" areas, where the oxygen diffuses and is the oxidising agent in the cathodic corrosion reaction. It is also observed that on both surfaces (**Figure 3c** and **d**), some particles remained (named as D and F), and according to EDS (**Table 3**), they correspond to cathodic rich in Fe [7], reported in the reference sample as α-Al(Fe,Mn,Cu)Si [35]; these intermetallic particles promoted the preferential dissolution of the aluminium matrix [40, 41].

3.2 OCP (free corrosion potential, φ_{corr}) measurements

Figure 4 compares the φ_{corr} fluctuations of AA6061-T6 specimens exposed for different periods of times (0, 5 and 30 days) under laminar flow conditions (**Figure 4a**) and without flow (**Figure 4b**). The average values are summarised in **Table 4**. The trends in the changes, towards more or less negative values, are a response to the transformations that have occurred on the surface of the aluminium alloy with the advance of the corrosion process. These are in the morphology, elemental composition of the layers formed, as well as the type of localised corrosion attack discussed previously. It can be seen (**Table 4**) that the initial values of φ_{corr} are relatively similar, being 30 mV nobler in respect of the exposed surface in laminar flow. At 5 days, which implies the initial destruction of the passive layer (**Figure 2a**), the φ_{corr} is 100 mV more negative with laminar flow than under

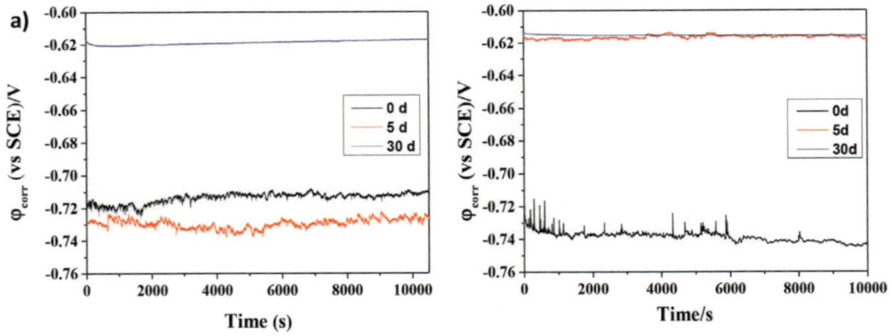

Figure 4.
Free corrosion potential (φ_{corr}) values of AA6061-T6 samples in Caribbean seawater during 30 days: (a) with laminar and (b) stationary flows.

Exposure time/days	φ_{corr} vs. SCE/mV (laminar flow)	φ_{corr} vs. SCE/mV (stationary flow)
0	-713 ± 2.80	-741 ± 16.40
5	-729 ± 2.70	-619 ± 0.95
30	-618 ± 1.00	-614 ± 0.28

Table 4.
Average of free corrosion potential (φ_{corr}) values for AA6061-T6 in Caribbean seawater under laminar and stationary flows at different times.

stationary flow. At 30 days, the φ_{corr} values in both seawater conditions (with and without laminar flow) are very similar (**Table 4**). However, when the product layer was removed (**Figure 3c** and **d**), the SEM images revealed a greater localised attack on the aluminium surface under laminar flow (**Figure 3c**), while in the absence of flow, this attack has been less aggressive, presenting shallow pits. Free corrosion potential (φ_{corr}) tendency towards more or less negative values indicates periods of corrosion activation or repassivation of the surface, both facts related to the characteristics of the layers formed [4, 42].

3.3 Surface characterisation by XPS

In order to identify the composition of the corrosion product layer created on the alloy surface after 30 days of exposure in laminar flow, XPS analysis was carried out on the specimen immersed, taking into account that aluminium corrosion products were not provided by XRD analysis as crystalline phases and are possibly amorphous.

Figure 5 shows the full XPS spectrum of the corrosion products formed on the aluminium alloy surface of 6061-T6. The XPS spectrum revealed signals of Mg, Na, O, Cl, C, Si and Al, which accord with EDS analysis. The high-resolution peak for Al2p, situated at 74.38 eV, has been associated with the presence of aluminium hydroxide [$Al(OH)_3$] [43], possibly derived from the transformation of basic aluminium chloride ($AlCl_3 \cdot H_2O$) [38]. On the other hand, the signal of O1s centred in 531.88 eV could be attributed to aluminium oxide (Al_2O_3), an important product of the repassivation process of the aluminium surface.

3.4 Electrochemical noise measurement

Figure 6 shows the current oscillations, and it can be seen that at the beginning of the experiment (0 days), the current density of AA6061-T6 surface immersed in

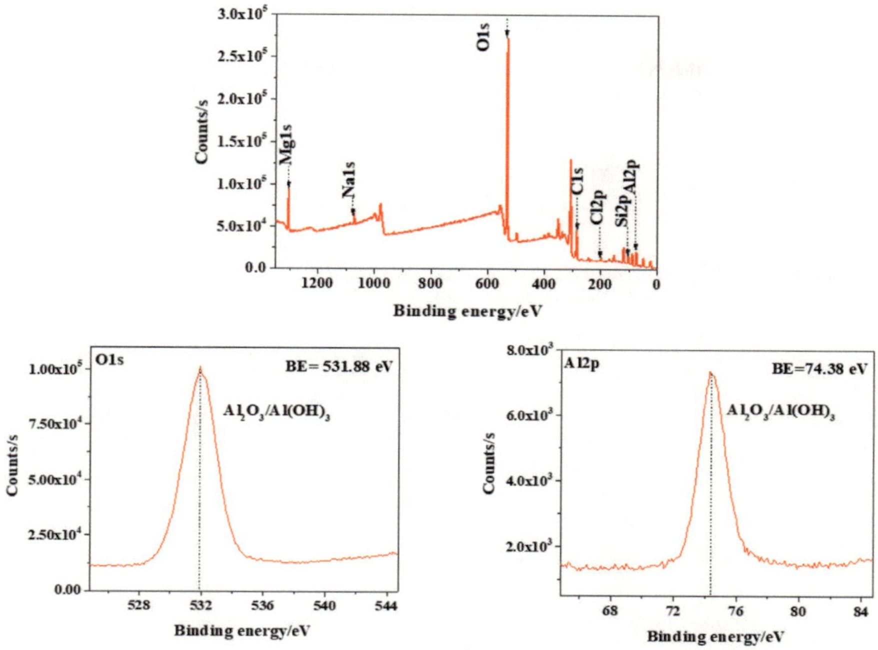

Figure 5.
Overview XPS spectra acquired from AA6061-T6 after 30 days of immersion in Caribbean seawater with laminar flow.

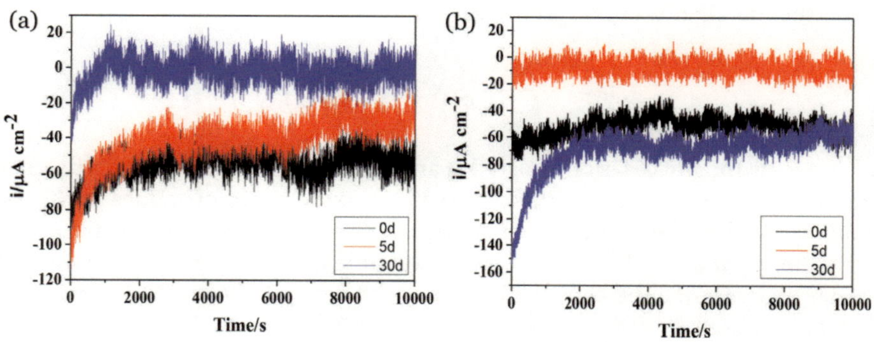

Figure 6.
Current density fluctuation for AA6061-T3 immersed in Caribbean seawater up to 30 days under (a) laminar and (b) stationary flows.

laminar flow of seawater is higher in ≈50 μA cm^{-2} than that current in stationary flow. This suggested that the corrosion of the aluminium alloy 6061-T6 surface in laminar flow initiates faster, when the oxide layer on the alloy begins to break down. However, at the end of the experiment (30 days), the current value diminished suddenly, compared with the initial values. However, in stationary flow the current shifted to one order higher values than those in laminar flow, suggesting an acceleration of the corrosion process at that period of time [44]. Conversely, the current oscillations in stationary conditions (**Figure 6b**) presented slow variations, while for flow conditions (**Figure 6a**), intense current fluctuations were acquired with greater amplitude, which suggest greater corrosion [45]. The observed current oscillations registered in (μA cm^{-2}) (**Figure 6**) correspond to the variation of the free corrosion potential (φ_{corr}) values in several mV (**Figure 4**).

The corrosion current density was calculated from the value of polarisation resistance R_p (on the assumption that R_p is equivalent to polarisation resistance

	R_p/kΩ cm^2		i_{corr} /μA cm^{-2}	
Time/d	**Laminar flow**	**Stationary flow**	**Laminar flow**	**Stationary flow**
0	3.65	23.46	7.12	1.11
5	2.24	17.5	11.60	1.49
30	1.44	19.3	18.08	1.35

Table 5.
Polarisation resistance (R_p) and corrosion current density values for AA6061-T6 immersed up to 30 days in Caribbean seawater with laminar and stationary flows.

	CR/mm year^{-1}	
Time/d	**Laminar flow**	**Stationary flow**
0	0.08	0.012
5	0.13	0.016
30	0.20	0.014

Table 6.
Corrosion rate of AA6061-T6 in Caribbean seawater from the Cozumel Channel under laminar and stationary flows.

noise R_n) obtained by ECN tests, according to the Stern-Geary equation (Eq. (5)). R_n is calculated by dividing the standard deviation of potential by standard deviation of current (the potential noise can be modelled as the action of the current noise on the metal-solution impedance):

$$i_{corr} = \frac{B}{R_p} = \frac{1}{R_p} x \frac{b_a b_c}{2.303\,(b_a + b_c)} \tag{5}$$

where R_p is the polarisation resistance and b_a and b_c are the Tafel coefficients. In this research, the B value employed was 0.26 V, taking common values for b_a and b_c of aluminium alloys [46, 47]. The calculated values (**Table 5**) show that the corrosion current of AA6061-T6 increases with the time of exposure, being more than one order higher when the alloy is exposed under laminar flow of seawater, while in stationary conditions, it maintains almost similar values up to 30 days.

Subsequently, Faraday's law was applied to calculate the corrosion rate (CR, **Table 6**) in the following form:

$$CR = \frac{i_{corr} K E_w}{\rho A} \tag{6}$$

where E_w stands for the equivalent mass of AA6061-T6, I_{corr} is the corrosion density (A cm^{-2}), ρ is the metal density, K is a constant (3272 mm/A cm year) and A is the exposed specimen area (1 cm^2) [48].

The corrosion rate values presented in **Table 6** indicate that under laminar flow, the values varied 0.08 and 0.20 mm per year, while in stationary conditions, they were between 0.012 and 0.162 mm per year.

With the statistical data obtained from the corrosion current, the pitting index (PI) [19] was calculated in order to reveal AA606-T6 susceptibility to localised corrosion for the laminar and stationary:

$$PI = \sigma_i (i_{rms})^{-1} = (639.17\,nA)(666.42\,nA)^{-1} = 0.96 \tag{7}$$

where σ_i is the standard deviation and i_{rms} the main square root of current noise.

Values of PI above 0.1 may indicate localised corrosion [20, 49]. The pitting indexes are shown in **Table 7**. Thus, the calculated PI value suggests that at the end of the experiment (30 days), for both flow cases, AA606-T6 showed pitting corrosion, approximately four times higher in flow conditions, reaching PI = 0.96. These facts agree with the SEM images (**Figure 3**) comparing the corrosion attacks on AA6061-T6 exposed to both flow conditions.

The current fluctuations, considered as EN, were transformed into the frequency domain to estimate PSD slopes (β exponent). **Figure 7** compares the PSD plots in bi-logarithmic scale, corresponding to AA6061-T6 surfaces after 30 days of exposure in seawater under laminar and stationary flows. In each case, the β exponent decreases as the frequency increases, and this fact could be associated with the advance of the localised corrosion attacks on the alloy surface [34]. At 30 days, β values are similar in laminar and stationary flows (1.0 and 0.94, respectively) and may be attributed to the fractional Gaussian noise (*fGn*), associated with a persistent process [26]. This type of noise (*fGn*) is considered also as a stationary process [24].

Figure 8 shows the spontaneous energy E during the corrosion process. At the beginning, after 5 days the energy was of an order of magnitude higher in laminar flow (1.1×10^{-4}). This fact is consistent with the SEM image presented in **Figure 2**, which showed the accelerated corrosion process causing severe damage to the alloy surface exposed to laminar flow.

However, at the end of the experiment (30 days), the energy diminished in magnitude, being very similar in the order of 10^{-7}, for both flows, probably on account of the formation of layers of corrosion products with different characteristics that act as a physical barrier on the alloy surfaces, slowing down the corrosion attack.

	σ_i/nA		i_{rms}/nA		Pitting index	
Time/days	**Laminar**	**Stationary**	**Laminar**	**Stationary**	**Laminar**	**Stationary**
0	772	639	5462	5255	0.14	0.12
5	1279	1146	4252	2866	0.30	0.40
30	639	1471	666	6733	0.96	0.22

Table 7.
Pitting index of AA6061-T6 immersed for 30 days in Caribbean seawater (Cozumel Channel) under laminar and stationary flows.

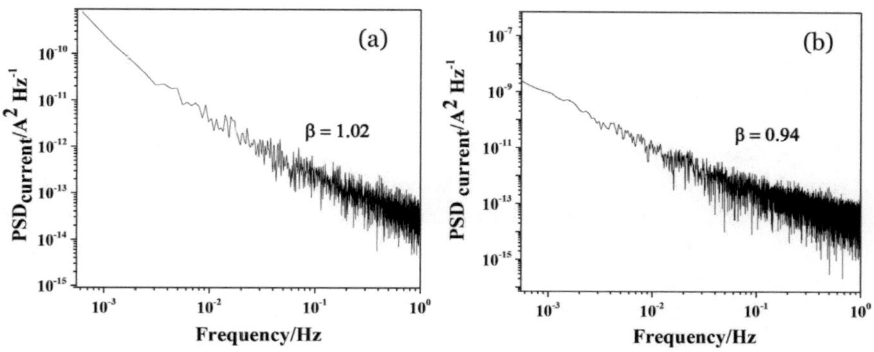

Figure 7.
Released energy (E) from AA6061-T6 surface after immersion in seawater under laminar flow, (a) at 5 days and (b) after 30 days of exposure; stationary flow, (c) at 5 days and (d) after 30 days of exposure.

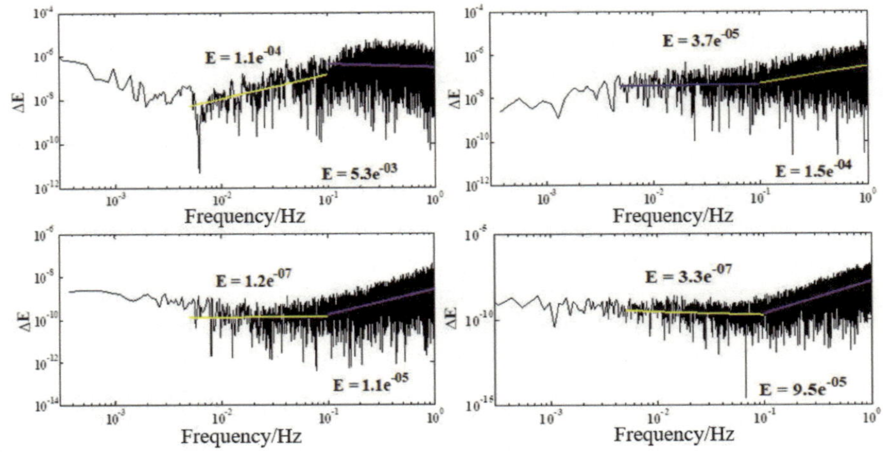

Figure 8.
Energy (E) from AA6061-T6 surface after immersion in seawater under laminar flow, (a) at 5 days and (b) after 30 days of exposure; stationary flow, (c) at 5 days and (d) after 30 days of exposure.

4. Conclusions

1. The initial electrochemical activity of 6061-T6 aluminium alloy surface, immersed in Caribbean seawater, was studied for 30 days under laminar flow (0.1 m s^{-1}, 21°C). The changes in the morphology and elemental composition of the formed layers have revealed localised corrosion (fissures and deep cracks) in the vicinity of intermetallic particles rich in Fe and Cu, which act as cathodes. The attack was less aggressive in stationary seawater, with shallow pits occurring on the surface at 30 days of exposure.

2. The calculated values showed that the corrosion current (icorr) of AA6061-T6 increases with the time of exposure more than one order higher when the alloy is exposed under laminar flow of seawater (7.12–18.08 μA cm^{-2}), while in stationary conditions, it maintains almost similar values (17.51–19.32 μA cm^{-2}). With the statistical data obtained from the corrosion current, the calculated pitting index (PI) revealed that AA606-T6 is four times more susceptible to localised corrosion in seawater under laminar flow (PI = 0.96), compared to that in stationary conditions (PI = 0.22). The estimated PSD slopes (β exponent) of the current fluctuations transformed into the frequency domain revealed that in laminar and stationary flows (β = 1.0 and 0.94, respectively), the electrochemical corrosion may be attributed to the fractional Gaussian noise (fGn), associated with a persistent stationary process.

3. The spontaneous energy release in the initial stages is one order higher in laminar flow ($\Delta E = 1.1 \times 10^{-4}$). However, at the end of the experiment (30 days), the energy diminished in magnitude, being very similar in an order of 10^{-7}, for both flows, probably on account of the formation of layers of corrosion products with different characteristic, which act as a physical barrier on the alloy surfaces, slowing down the corrosion attack.

4. The observed effect of the laminar seawater flow on the AA6061-T6 corrosion process should be considered as a consequence of the facilitated diffusion of the oxygen at the metal-seawater interface, resulting in specific transformation

in the composition of the formed corrosion layers. The presented surface SEM-EDS and XPS analysis agree positively with the results obtained with both nondestructive electrochemical methods.

Acknowledgements

The authors acknowledge LANNBIO for permitting the use of their facilities, as well as to M. Sci. Dora Huerta and W. J. Cauich-Ruiz for their technical assistance in data acquisition.

Funding sources

This research was funded by Centro Mexicano de Inovación en Energía del Océano (CEMIE) grant number [00249795]. Luis Chávez gratefully thanks CONACYT for his scholarship as M.Sci. student at CINVESTAV-IPN.

Author details

Gloria Acosta, Lucien Veleva*, Luis Chávez and Juan L. López
Applied Physics Department, Research Center for Advanced Study (CINVESTAV-IPN), Mérida, Yucatán, México

*Address all correspondence to: veleva@cinvestav.mx

References

[1] Davis JR. Understanding the Basics. Ohio: ASM International; 2001. 351 p. DOI: 10.1361/autb2001p351

[2] Altenpohl DG. Aluminum Viewed from Within: An Introduction into the Metallurgy of Aluminum Fabrication. Düsseldorf: Aluminum-Verlag; 1982

[3] Pourbaix M. Atlas of Electrochemical Equilibria in Aqueous Solutions. Huston, TX: NACE International; 1974

[4] Szklarska-Smialowska Z. Pitting corrosion of aluminum. Corrosion Science. 1999;**41**:1743-1767. DOI: 10.1016/S0010-938X(99)00012-8

[5] Tiryakioğlu M, Staley J. Physical metallurgy and the effect of alloying additions in aluminum alloys. In: Totten GE, MacKenzie DS, editors. Handbook of Aluminum. New York: Marcel Dekker, Inc; 2003. pp. 81-210

[6] Reboul MC, Baroux B. Metallurgical aspects of corrosion resistance of aluminium alloys. Materials and Corrosion. 2011;**62**:215-233. DOI: 10.1002/maco.201005650

[7] Mutombo K. Intermetallic particles-induced pitting corrosion in 6061-T651 aluminium alloy. Materials Science Forum. 2011;**690**:389-392. DOI: 10.4028/www.scientific.net/MSF.690.389

[8] Guillaumin V, Mankowski G. Localized corrosion of 2024 T351 aluminium alloy in chloride media. Corrosion Science. 1999;**41**:421-438. DOI: 10.1016/S0010-938X(98)00116-4

[9] Gharavi F, Matori K, Yunus R, Othman NK, Fadaeifard F. Corrosion evaluation of friction stir welded lap joints of AA6061-T6 aluminum alloy. Transactions of Nonferrous Metals Society of China. 2016;**26**:684-696. DOI: 10.1016/S1003-6326(16)64159-6

[10] Zheng Y, Luo B, Bai Z, Wang J, Yin Y. Study of the precipitation hardening behavior and intergranular corrosion of Al-Mg-Si alloy with differing Si contents. Meta. 2017;**7**: 387-399. DOI: 10.3390/met7100387

[11] F-l Z, Z-l W, Li J-f, Li C-x, Tan X, Zhang Z, et al. Corrosion mechanism associated with Mg_2Si and Si particles in Al−Mg−Si alloys. Transactions of Nonferrous Metals Society of China. 2011;**21**:2559-2567. DOI: 10.1016/S1003-6326(11)61092-3

[12] Birbilis N, Buchheit RG. Electrochemical characteristics of intermetallic phases in aluminum alloys an experimental survey and discussion. Journal of the Electrochemical Society. 2005;**152**:B140-B151. DOI: 10.1149/1.1869984

[13] McCafferty E. Sequence of steps in the pitting of aluminum by chloride ions. Corrosion Science. 2003;**5**: 1421-1438. DOI: 10.1016/S0010-938X(02)00231-7

[14] Burstein GT, Liu C, Souto RM, Vines SP. Origins of pitting corrosion. Corrosion Engineering, Science and Technology. 2004;**39**:25-30. DOI: 10.1179/147842204225016859

[15] Liang M, Melchers R, Chaves I. Corrosion and pitting of 6060 series aluminium after 2 years exposure in seawater splash, tidal and immersion zones. Corrosion Science. 2018;**140**: 286-296. DOI: 10.1016/j.corsci.2018.05.036

[16] LaQue FL. Marine Corrosion: Causes and Prevention. New York: John Wiley & Sons; 1975

[17] Al-Fozan SA, Malik AU. Effect of seawater level on corrosion behavior of different alloys. Desalination. 2008;**228**: 61-67. DOI: 10.1016/j.desal.2007.08.007

[18] Roberge PR. Handbook of Corrosion Engineering. New York: McGraw-Hill; 2000

[19] Dawson JL. Electrochemical noise measurement: The definitive in-situ technique for corrosion applications. In: Kearns J, Scully J, Roberge P, Reichert D, Dawson JL, editors. Electrochemical Noise Measurement for Corrosion Applications. West Conshohocken, PA: ASTM STP 1277; 1996. pp. 3-35. DOI: 10.1520/STP37949S

[20] Xia DH, Song SZ, Behnamian Y. Detection of corrosion degradation using electrochemical noise (EN): Review of signal processing methods for identifying corrosion forms. Corrosion Engineering, Science and Technology. 2016;**51**:527-544. DOI: 10.1179/1743278215Y.0000000057

[21] Casajús P, Winzer N. Electrochemical noise analysis of the corrosion of high-purity Mg–Al alloys. Corrosion Science. 2015;**94**:316-326. DOI: 10.20964/2018.01.86

[22] Lee CC, Mansfeld F. Analysis of electrochemical noise data for a passive system in the frequency domain. Corrosion Science. 1998;**40**:956-962. DOI: 10.1016/j.arabjc.2012.02.018

[23] Roberge PR. Quantifying the stochastic behavior of electrochemical noise measurement during the corrosion of aluminum. In: Kearns JR, Scully JR, Roberge PR, Reichert DL, Dawson JL, editors. Electrochemical Noise Measurement for Corrosion Applications. West Conshohocken, PA: ASTM STP 1277; 1996. pp. 142-156. DOI: 10.1520/STP37957S

[24] Eke A, Hermán P, Bassingthwaighte JB, Raymond GM, Percival DB, Cannon M, et al. Physiological time series: Distinguishing fractal noises from motions. European Journal of Physiology. 2000;**439**: 403-415. DOI: 10.1007/s004249900135

[25] Delignieres D, Ramdani S, Lemoine L, Torre K, Fortes M, Ninot G. Fractal analyses for 'short' time series: A re-assessment of classical methods. Journal of Mathematical Psychology. 2006;**50**:525-544. DOI: 10.1016/j.jmp.2006.07.004

[26] Planinšič P, Petek A. Characterization of corrosion processes by current noise wavelet-based fractal and correlation analysis. Electrochimica Acta. 2008;**53**:5206-5214. DOI: 10.1016/j.jmp.2006.07.004

[27] López JL, Veleva L, López-Sauri DA. Multifractal detrended analysis of the corrosion potential fluctuations during copper patina formation on its first stages in sea water. International Journal of Electrochemical Science. 2014;**9**: 1637-1649

[28] Acosta G, Veleva L, López JL. Power spectral density analysis of the corrosion potential fluctuation of aluminium in early stages of exposure to Caribbean sea water. International Journal of Electrochemical Science. 2014;**9**: 6464-6474

[29] Acosta G, Veleva L, López JL, López-Sauri DA. Contrasting initial events of localized corrosion on surfaces of 2219-T42 and 6061-T6 aluminum alloys exposed in Caribbean seawater. Transactions of Nonferrous Metals Society of China. 2019;**29**:34-42

[30] Mena-Morcillo E, Veleva LP, Wipf DO. Multi-scale monitoring the first stages of electrochemical behavior of AZ31B magnesium alloy in simulated body fluid. Journal of the Electrochemical Society. 2018; **165**:C749-C755. DOI: 10.1149/2.0291811jes

[31] McCabe WL, Smith JC, Harriott P. Unit Operations of Chemical Engineering. 4th ed. New York: McGraw Hill; 1985

[32] ASTM G1−90. Standard practice for preparing, cleaning, and evaluation corrosion test specimens

[33] ASTM G199-09. Standard guide for electrochemical noise measurement

[34] Stein JY. Digital Signal Processing: A Computer Science Perspective. 2nd ed. New York: John Wiley & Sons; 2000

[35] Yasakau KA, Zheludkevich ML, Ferreira MGS. Role of intermetallics in corrosion of aluminum alloys. In: Mitra R, editor. Smart Corrosion Protection. Aveiro, Portugal: Woodhead Publishing; 2018. 425 p. DOI: 10.1016/B978-0-85709-346-2.00015-7

[36] Ender VV, Wetzel C. Gross-Kraftwerkbret: VGB Tech. Ver; 1998, R 6/1-R6/20

[37] Afzal SN, Shaikh MA, Mustafa CM, Nabi M, Ehsan MQ, Khan AH. Study of aluminum corrosion in chloride and nitrate media and its inhibition by nitrite. Journal of Nepal Chemical Society. 2006;**22**:26-33. DOI: 10.3126/jncs.v22i0.519

[38] Foley RT, Nguyen TH. The chemical nature in aluminum corrosion, V. energy transfer in aluminum dissolution. Journal of the Electrochemical Society. 1982;**129**: 464-467. DOI: 10.1149/1.2123881

[39] Brown RR, Daut GE, Mrazek RV, Gokcen NA. Solubility and Activity of Aluminum Chloride in Aqueous Hydrochloric Acid Solutions. Washington, DC: U.S. Department of the Interior, Bureau of Mines; 1979

[40] Nikseresht Z, Karimzadeh F, Golozar MA, Heidarbeigy M. Effect of heat treatment on microstructure and corrosion behavior of Al6061 alloy weldment. Materials and Design. 2010; **31**:2643-2648. DOI: 10.1016/j.matdes.2009.12.001

[41] Zhu Y, Sun K, Frankel GS. Intermetallic phases in aluminum alloys and their roles in localized corrosion. Journal of the Electrochemical Society. 2018;**165**:C807. DOI: 10.1149/2.0931811jes

[42] Ghali E. General, galvanic, and localized corrosion of aluminum and its alloys. In: Winston Revie R, editor. Corrosion Resistance of Aluminum and Magnesium Alloys: Understanding, Performance, and Testing. Hoboken, New Jersey: John Wiley & Sons; 2010. 176 p. DOI: 10.1002/9780470531778.ch5

[43] John FM, William FS, Peter ES, Kenneth DB. In: Chastain J, editor. Handbook of X-ray Photoelectron Spectroscopy. Minnesota: Physical Electronics; 1992. pp. 44-55

[44] Loto CA. Electrochemical noise measurement technique in corrosion research. International Journal of Electrochemical Science. 2012;**7**: 9248-9270

[45] Malo JM, Velazco O. Elecrochemical noise response of steel under hydrodynamic conditions. In: Kearns JR, Scully JR, Roberge PR, Reichert DL, Dawson JL, editors. Electrochemical Noise Measurement for Corrosion Applications. West Conshohocken, PA: ASTM STP 1277; 1996. pp. 387-397. DOI: 10.1520/STP37972S

[46] Sanchez-Amaya JM, Cottis RA, Botana FJ. Shot noise and statistical parameters for the estimation of corrosion mechanisms. Corrosion Science. 2005;**47**:3280-3299. DOI: 10.1016/j.corsci.2005.05.047

[47] Curioni M, Cottis RA, Thompson GE. Application of electrochemical noise analysis to corroding aluminium alloys. Surface and Interface Analysis. 2013;**45**:1564-1569. DOI: 10.1002/sia.5173

[48] G102-89. Standard practice for calculation of corrosion rates and related information from electrochemical measurements

[49] Eden DA, John DG, Dawson JL. Corrosion monitoring: WIPO Patent, 1987, 19871107022. 1987. pp. 11–19

Chapter 9

A Unique Volume Balance Approach for Verifying the Three-Dimensional Hydrodynamic Numerical Models in Surface Waterbody Simulation

Hussein A.M. Al-Zubaidi and Scott A. Wells

Abstract

The hydrodynamic numerical modeling is increasingly becoming a widely used tool for simulating the surface waterbodies including rivers, lakes, and reservoirs. A challenging step in any model development is the verification tests, especially at the early stage of development. In this study, a unique approach was developed by implementing the volume balance principle in order to verify the three-dimensional hydrodynamic models for surface waterbody simulation. A developed and verified three-dimensional hydrodynamic and water quality model, called W3, was employed by setting a case study model to be verified using the volume balance technique. The model was qualified by calculating the error in the accumulated water volume within the domain every time step. Results showed that the volume balance reached a constant error over the simulation period, indicating a robust model setup.

Keywords: hydrodynamic model, lakes and reservoirs, model verification, model simulation, numerical model, volume balance, water quality modeling, W3 model

1. Introduction

Many 3D hydrodynamic and water quality models have been developed since the 1960s, and different numerical solution techniques have been used to solve the governing equations. The most popular numerical models and the basis that other models has been built based on are POM [1, 2], ECOM [1, 3], NCOM [4, 5], FVCOM [6, 7], EFDC [8], TRIM-3D [9], UnTRIM [10], GLLVHT [11], and DNS [12].

During the development stage of any numerical model, verification tests need to be performed to ensure that model foundations are valid. The 3D simulation models available in market have been tested either by comparing the predictions with the analytical solution, field data, or both. As a result, each verification approach has its advantages and disadvantages depending on the model complexity (governing equations used to develop the model and assumptions used to simplify the problem).

All three-dimensional models available to simulate surface waterbodies do not have outputs related to the model of volume balance performance (see the user manuals of the above popular models). Therefore, the user does not know the model preserves volume or not during the simulation period even though the model gives results. In addition, most 3D users run the simulation for a very short time (even for seconds), thinking the model is stable, since the 3D numerical models require long computation time to run. Thus, the need to develop a new volume balance tool arises based on these issues related to 3D hydrodynamic numerical models used in practice for surface waterbodies.

In this work, the volume balance approach was used as a tool to measure how a model preserves volume during the simulation time by calculating the accumulated error over time as a percent. Therefore, the modeler can monitor the model performance over time and decide whether the model is robust or not while running the model rather than waiting until the end of simulation.

2. Methods

To implement the volume balance approach, the three-dimensional model W3 developed by [13] for modeling hydrodynamics, temperature, and water quality in surface waterbodies was employed. Using the finite differences, the model solves the governing equations of continuity, free surface, momentums, and mass transport. Comparisons with analytical solutions and field data were carried out for verifying and validating the W3 model [13–17].

The model of volume balance was performed by comparing the water volume in the model domain during a time period with the water volume entering and leaving the same domain during the same period of time.

Let Vol be the accumulated water volume in the model domain over time. Then,

$$\mathrm{Vol} = \mathrm{Vol}_{\mathrm{initial}} + \mathrm{Vol}_{\mathrm{in}} - \mathrm{Vol}_{\mathrm{out}} \tag{1}$$

where $\mathrm{Vol}_{\mathrm{initial}}$ = the initial water volume within the domain; $\mathrm{Vol}_{\mathrm{in}}$ = the accumulated water volume entering the domain; and $\mathrm{Vol}_{\mathrm{out}}$ = the accumulated water volume leaving the domain.

Thus, the error over time can be calculated as follows:

$$\%\mathrm{Error} = \left(\mathrm{abs}\left(\mathrm{Vol} - \mathrm{Vol}_{\mathrm{internal}}\right)\right)/\mathrm{Vol}_{\mathrm{internal}} \times 100 \tag{2}$$

where $\mathrm{Vol}_{\mathrm{internal}}$ is the water volume within the domain at any time during the simulation period.

A subroutine was added to the model to check the volume preservation by calculating % error at every time step. A lower % error represents more accurate model predictions. The error should reach a constant value with time and should not grow with time. If % error grows with time exponentially, this implies that the model goes unstable (blows up). Two tests implementing the volume balance check were performed. One of these tests examined the volume balance over a rectangular domain, and the other tests evaluated the volume balance over an irregular domain. Both tests were performed over a period of 100 days based on the same real meteorological data, calculated solar short radiation, and constant inflow and outflow. The meteorological data are shown in **Figures 1–5**.

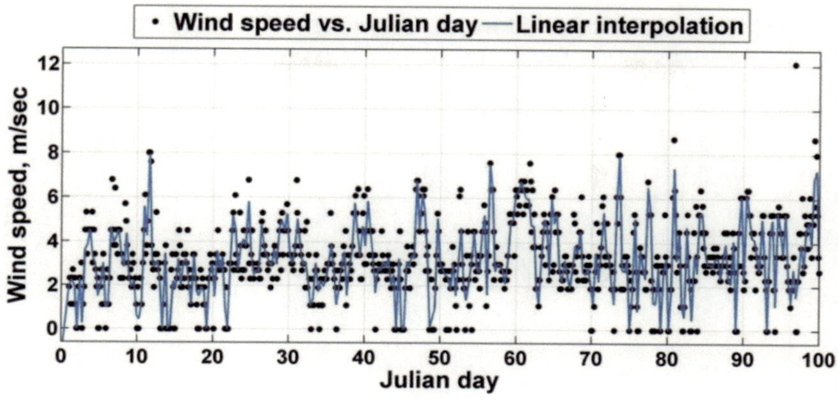

Figure 1.
Wind speed input data.

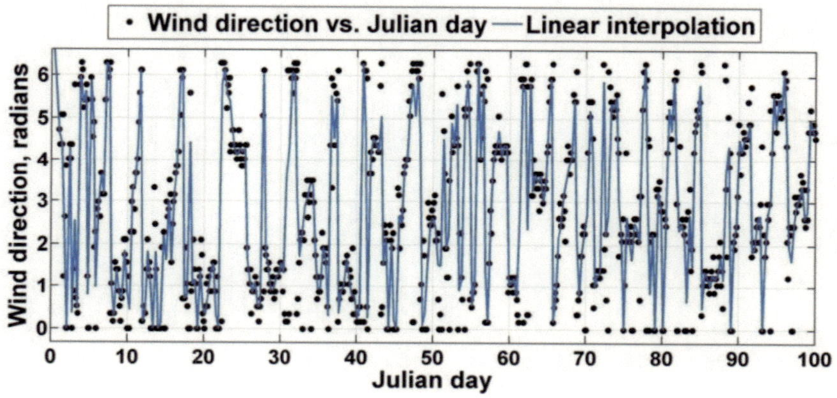

Figure 2.
Wind direction input data.

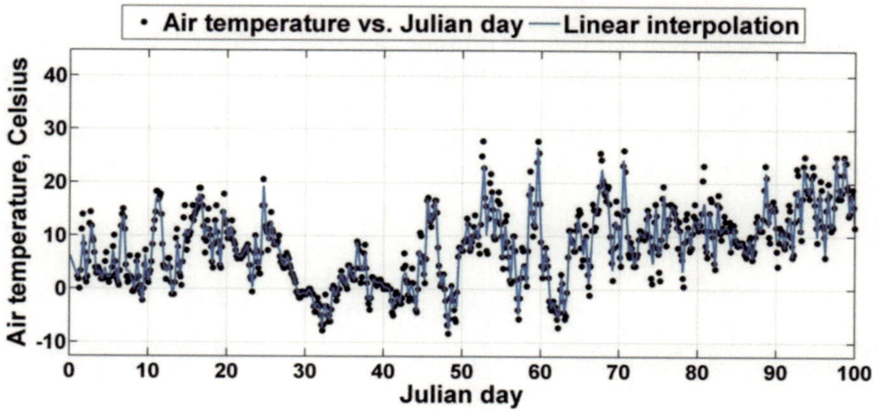

Figure 3.
Air temperature input data.

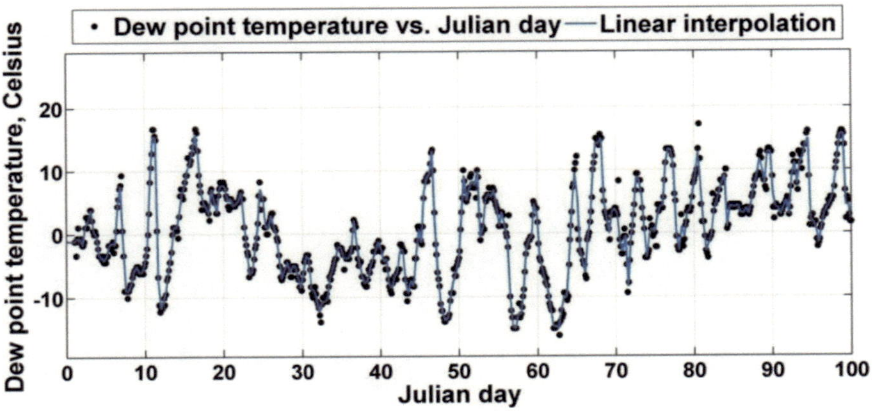

Figure 4.
Dew point input data.

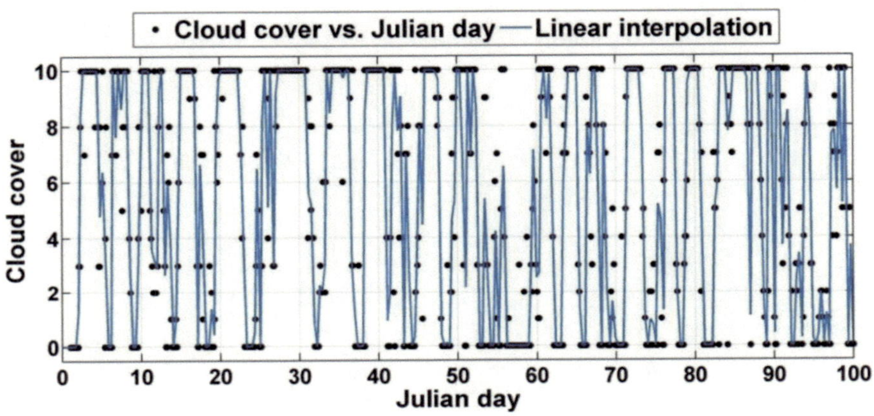

Figure 5.
Cloud cover input data.

3. Results and discussion

The physical domain was divided into computational cells of 1000 × 500 × 1 (x,y,z) m and oriented perpendicular to the north direction as shown in **Figure 6**, in which there are bends at the boundaries to check how the model catches the flow field variability. The code was run without assuming a frictionless fluid, with the Coriolis force, with wind variable in magnitude and direction at 10 m height above

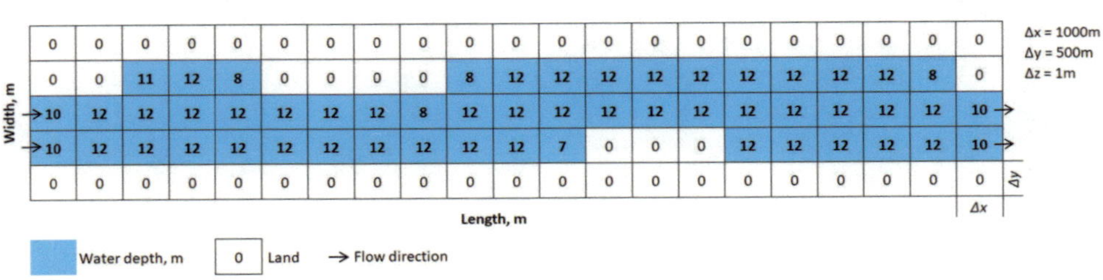

Figure 6.
Irregular physical domain and the input bathymetry.

the water surface, with a constant inflow and outflow of 0.8 m^3/s, and with variable water temperature over time by solving the heat transport equation. Additionally, the adding/subtracting layers algorithm (see [18]) was turned on to examine the surface layer thickness over the simulation period.

Using a time step of 35 s and a degree of implicitness (θ) of 1, the code was run for the simulation period. **Figure 7** presents the model predictions of the surface velocity field at Julian day 100. The model results showed good performance in following the bends at the boundaries. Furthermore, the volume balance error gave a good agreement in preserving volume in which the percent error reached a constant low value over time as shown in **Figure 8**, which is a semilog plot of the percent error with time. The corresponding water levels at three locations over time were shown in **Figure 9**, denoting a very small change (≅0.005 m) in the surface layer thickness resulting from the free water surface waves.

Since the W3 model uses the degree of implicitness to switch between the fully implicit numerical scheme and the fully explicit scheme, the effect of the degree of implicitness on the accumulated error was evaluated by running the code using θ = 0.5 with the same inputs that were used with θ = 1. The results showed that using the semi-implicit scheme of θ = 0.5 produces less percent error than using θ = 1. **Figure 10** shows the percent error after running the code for day 100 using two degrees of implicitness (θ = 1 and θ = 0.5).

In addition and in order to make sure that the numerical answers do not depend on the grid resolution, a grid refinement was performed, and the associated volume

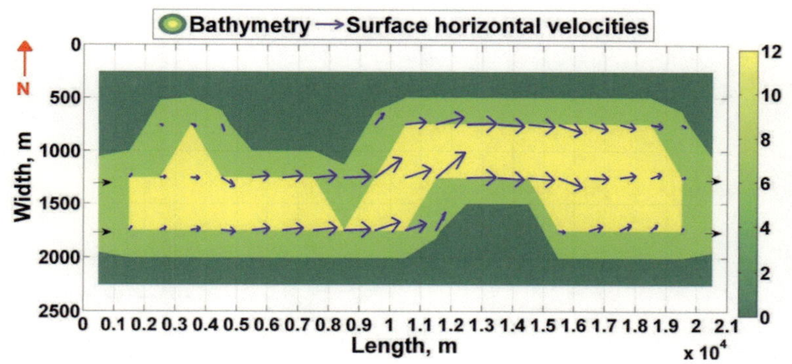

Figure 7.
Surface velocity field for the irregular domain at Julian day 100.

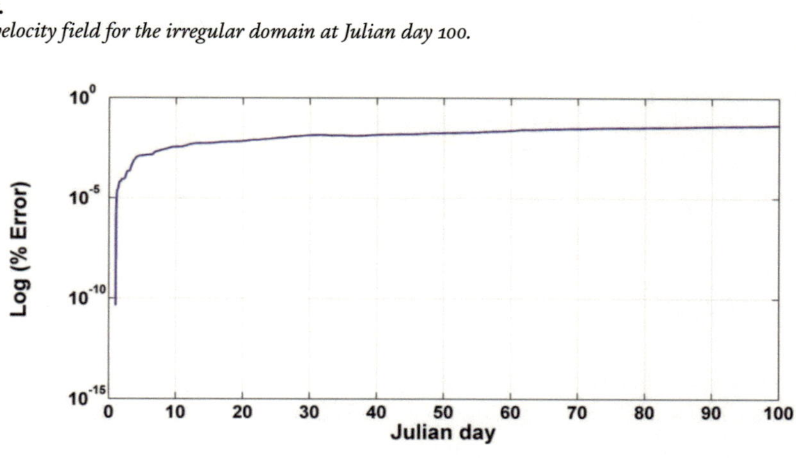

Figure 8.
Volume balance for the irregular domain using θ = 1.

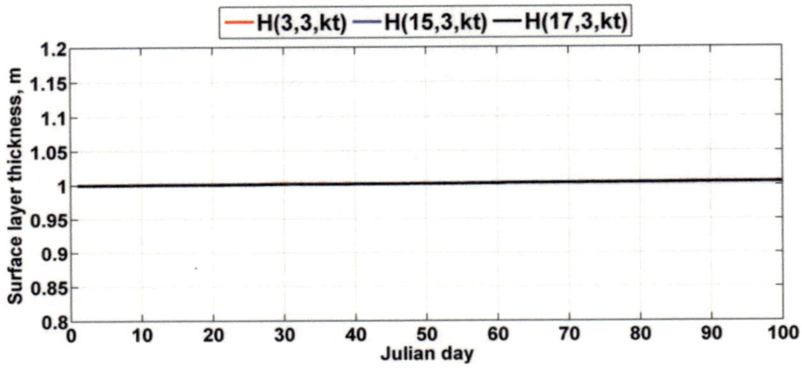

Figure 9.
Surface layer thickness over time for the irregular domain using $\theta = 1$.

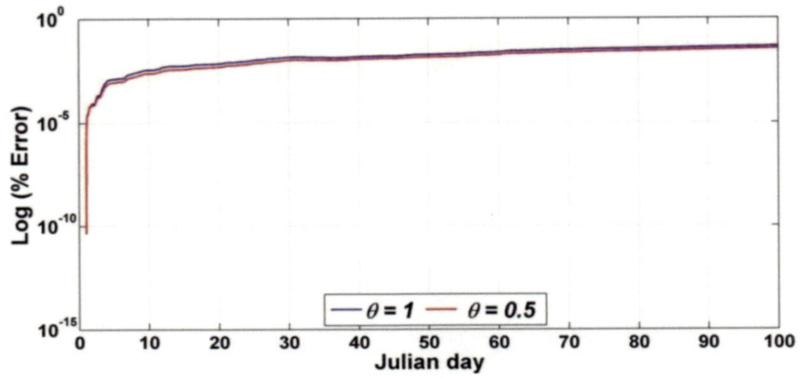

Figure 10.
The volume balance for the irregular domain using $\theta = 1$ and $\theta = 0.5$.

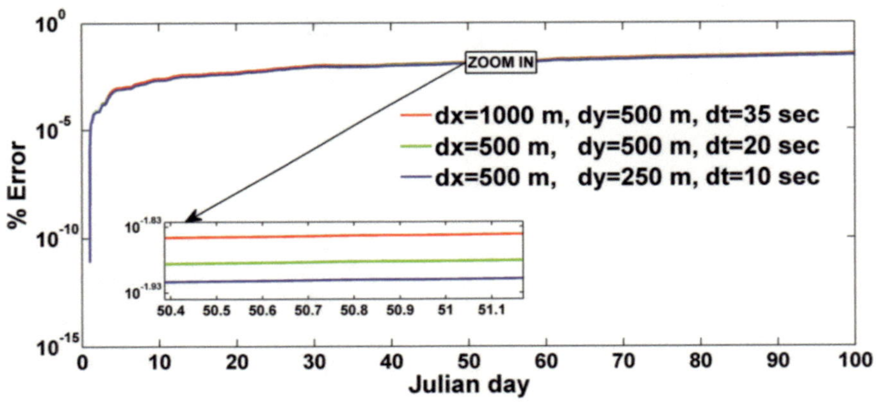

Figure 11.
The effect of grid refinement.

error was assessed. The code was run using $\theta = 0.5$ with three horizontal grid resolutions 1000 × 500, 500 × 500, and 500 × 125 (x,y) m in which the model was stable numerically. To maintain the stability, three different time steps were chosen to run the code because the refinement lowers the time step (Δt). All resolutions were applied on the same initial water volume in **Figure 6**. Therefore, the initial water volume of the waterbody was fixed, while the grid resolution was varied.

Figure 11 shows the percent error over time for the three considered grid resolutions, indicating that the error in volume has the same order of magnitude for the three resolutions.

4. Conclusions

Model verification is the first step after building any new hydrodynamic numerical model for surface waterbody simulation. In this chapter, a new volume balance approach was introduced for verifying the three-dimensional hydrodynamic numerical models in surface waterbody simulation. This technique provides information about whether the code preserves fluid mass or not by calculating the volume balance percent error over time during a model simulation. The model results indicated that the model is considered numerically stable if the volume balance error reaches a constant value over time. In addition, even though the model degree of implicitness had a reasonable volume balance error (less than 0.1%), the semi-implicit numerical scheme had slightly better volume balance error than the fully implicit scheme.

Acknowledgements

The authors thank the Department of Civil and Environmental Engineering, Portland State University, Portland, OR, USA, for their help in doing this research in association with the Iraqi Ministry of Higher Education and Scientific Research, University of Babylon.

Author details

Hussein A.M. Al-Zubaidi[1,2]* and Scott A. Wells[2]

1 Department of Environmental Engineering, College of Engineering, University of Babylon, Babylon, Iraq

2 Department of Civil and Environmental Engineering, Portland State University, Portland, OR, USA

*Address all correspondence to: alzubaidih10@gmail.com; hmahdi@pdx.edu

References

[1] Blumberg AF, Mellor GL. A description of a three-dimensional coastal ocean circulation model. In: Heaps NS, editor. Three-Dimensional Coastal Ocean Models. Washington, DC: American Geophysical Union; 1987. pp. 1-16. DOI: 10.1029/co004p0001

[2] Mellor GL. Users Guide for a Three-Dimensional, Primitive Equation, Numerical Ocean Model (October 2002 Version). Princeton, NJ: Program in Atmospheric and Oceanic Sciences, Princeton University; 2002

[3] Blumberg AF. A coastal ocean numerical model. In: Mellor GL, editor. Proceedings of International Symposium on Mathematical Modelling of Estuarine Physics. Berlin: Springer; 1980. pp. 202-219

[4] Barron CN, Kara AB, Martin PJ, Rhodes RC, Smedstad LF. Formulation, implementation and examination of vertical coordinate choices in the Global Navy Coastal Ocean Model (NCOM). Ocean Modelling. 2006;**11**(3-4):347-375. DOI: 10.1016/j.ocemod.2005.01.004

[5] Martin PJ, Barron CN, Smedstad LF, Campbell AJ, Rhodes RC, Rowley C, et al. User's Manual for the Navy Coastal Ocean Model (NCOM) Version 4.0. MS: Naval Research Laboratory, Oceanography Division, Stennis Space Center; 2009. Report No. NRL/MR/7320-08-9151

[6] Chen C, Beardsley RC, Cowles G, Qi J, Lai Z, Gao G, et al. An Unstructured-Grid, Finite-Volume Community Ocean Model FVCOM User Manual. 3rd ed. Cambridge, MA: Massachusetts Institute of Technology; 2011. Report No. 11-1101

[7] Chen C, Liu H, Beardsley RC. An unstructured grid, finite-volume, three-dimensional, primitive equations ocean model: Application to coastal ocean and estuaries. Journal of Atmospheric and Oceanic Technology. 2003;**20**:159-186. DOI: 10.1175/1520-0426(2003)020<0159:AUGFVT>2.0.CO;2

[8] Hamrick JM. A three-dimensional environmental fluid dynamics computer code: Theoretical and computational aspects. In: Special Report 317 in Applied Marine Science and Ocean Engineering. VA: Virginia Institute of Marine Science, School of Marine Science, College of William and Mary; 1992

[9] Casulli V, Cheng RT. Semi-implicit finite difference methods for three-dimensional shallow water flow. International Journal for Numerical Methods in Fluids. 1992;**15**(6):629-648. DOI: 10.1002/fld.1650150602

[10] Casulli V, Walters RA. An unstructured grid, three dimensional model based on the shallow water equations. International Journal for Numerical Methods in Fluids. 2000;**32**:331-348. DOI: 10.1002/(SICI)10970363(20000215)32:3<331::AID-FLD941>3.0.CO;2-C

[11] Edinger JE. Waterbody Hydrodynamic and Water Quality Modeling: An Introductory Workbook and CD-ROM on Three-Dimensional Waterbody Modeling. VA: ASCE; 2001

[12] Moin P, Mahesh K. Direct numerical simulation: A tool in turbulence research. Annual Review of Fluid Mechanics. 1998;**30**(1):539-578. DOI: 10.1146/annurev.fluid.30.1.539

[13] Al-Zubaidi HAM, Wells SA. Analytical and field verification of a 3D hydrodynamic and water quality numerical scheme based on the 2D formulation in CE-QUAL-W2. Journal of Hydraulic Research. 2018. DOI: 10.1080/00221686.2018.1499051. https://www.tandfonline.com/doi/full/10.1080/00221686.2018.1499051

[14] Al-Zubaidi HAM, Wells SA. 3D numerical temperature model development and calibration for lakes and reservoirs: A case study. In: Proceedings of the World Environmental and Water Resources Congress 2017. Sacramento, CA: ASCE; 2017. DOI: 10.1061/9780784480601.051

[15] Al-Zubaidi HAM, Wells SA. 2D and 3D numerical modeling of water level and temperature in lakes and reservoirs based on the numerical scheme in CE-QUAL-W2: A case study. In: Proceedings, the 37th International Symposium of the North American Lakes Management Society. Westminster, Colorado: NALMS; 2017

[16] Al-Zubaidi HAM, Wells SA. Comparison of a 2D and 3D hydrodynamic and water quality model for lake systems. In: Proceedings, World Environmental and Water Resources Congress 2018. Minneapolis, Minnesota: ASCE; 2018. DOI: 10.1061/9780784481400.007

[17] Al-Zubaidi HAM, Wells SA. Water level, temperature, and water quality numerical predictions of a 3D semi-implicit scheme for lakes and reservoirs: An analytical and field case study. In: Proceedings, the 9th International Congress on Environmental Modelling and Software (iEMSs 2018). Ft. Collins, Colorado: iEMSs; 2018

[18] Cole T, Wells SA. CE-QUAL-W2: A Two-Dimensional, Laterally Averaged, Hydrodynamic and Water Quality Model (Version 4.1). Portland, OR: Department of Civil and Environmental Engineering, Portland State University; 2017. Available from: http://www.cee.pdx.edu/w2/